Confidence Intervals in Generalized Regression Models

Recent Titles

Visualizing Statistical Models and Concepts, *R. W. Farebrother and Michaël Schyns*

Financial and Actuarial Statistics: An Introduction, *Dale S. Borowiak*

Nonparametric Statistical Inference, Fourth Edition, Revised and Expanded, *Jean Dickinson Gibbons and Subhabrata Chakraborti*

Computer-Aided Econometrics, *edited by David E.A. Giles*

The EM Algorithm and Related Statistical Models, *edited by Michiko Watanabe and Kazunori Yamaguchi*

Multivariate Statistical Analysis, Second Edition, Revised and Expanded, *Narayan C. Giri*

Computational Methods in Statistics and Econometrics, *Hisashi Tanizaki*

Applied Sequential Methodologies: Real-World Examples with Data Analysis, *edited by Nitis Mukhopadhyay, Sujay Datta, and Saibal Chattopadhyay*

Handbook of Beta Distribution and Its Applications, *edited by Arjun K. Gupta and Saralees Nadarajah*

Item Response Theory: Parameter Estimation Techniques, Second Edition, *edited by Frank B. Baker and Seock-Ho Kim*

Statistical Methods in Computer Security, *edited by William W. S. Chen*

Elementary Statistical Quality Control, Second Edition, *John T. Burr*

Data Analysis of Asymmetric Structures, *Takayuki Saito and Hiroshi Yadohisa*

Mathematical Statistics with Applications, *Asha Seth Kapadia, Wenyaw Chan, and Lemuel Moyé*

Advances on Models, Characterizations and Applications, *N. Balakrishnan, I. G. Bairamov, and O. L. Gebizlioglu*

Survey Sampling: Theory and Methods, Second Edition, *Arijit Chaudhuri and Horst Stenger*

Statistical Design of Experiments with Engineering Applications, *Kamel Rekab and Muzaffar Shaikh*

Quality by Experimental Design, Third Edition, *Thomas B. Barker*

Handbook of Parallel Computing and Statistics, *Erricos John Kontoghiorghes*

Statistical Inference Based on Divergence Measures, *Leandro Pardo*

A Kalman Filter Primer, *Randy Eubank*

Introductory Statistical Inference, *Nitis Mukhopadhyay*

Handbook of Statistical Distributions with Applications, *K. Krishnamoorthy*

A Course on Queueing Models, *Joti Lal Jain, Sri Gopal Mohanty, and Walter Böhm*

Univariate and Multivariate General Linear Models: Theory and Applications with SAS, Second Edition, *Kevin Kim and Neil Timm*

Randomization Tests, Fourth Edition, *Eugene S. Edgington and Patrick Onghena*

Design and Analysis of Experiments: Classical and Regression Approaches with SAS, *Leonard C. Onyiah*

Analytical Methods for Risk Management: A Systems Engineering Perspective, *Paul R. Garvey*

Confidence Intervals in Generalized Regression Models, *Esa Uusipaikka*

Confidence Intervals in Generalized Regression Models

Esa Uusipaikka

University of Turku

Turun Yliopisto, Finland

CRC Press
Taylor & Francis Group
Boca Raton London New York

CRC Press is an imprint of the
Taylor & Francis Group, an **informa** business

CRC Press
Taylor & Francis Group
6000 Broken Sound Parkway NW, Suite 300
Boca Raton, FL 33487-2742

First issued in paperback 2019

© 2009 by Taylor & Francis Group, LLC
CRC Press is an imprint of Taylor & Francis Group, an Informa business

No claim to original U.S. Government works

ISBN-13: 978-1-4200-6027-0 (hbk)
ISBN-13: 978-0-367-38708-2 (pbk)

Library of Congress Cataloging-in-Publication Data

Uusipaikka, Esa I.
 Confidence intervals in generalized regression models / Esa I. Uusipaikka.
  p. cm. -- (Statistics, textbooks and monographs ; 194)
 Includes bibliographical references and index.
 ISBN 978-1-4200-6027-0 (alk. paper)
 1. Regression analysis. 2. Confidence intervals. 3. Linear models (Mathematics) 4.
Statistics. I. Title. II. Series.

QA278.2.U97 2008
519.5'36--dc22   2008013345

Visit the Taylor & Francis Web site at
http://www.taylorandfrancis.com

and the CRC Press Web site at
http://www.crcpress.com

To
My children Silja, Lenni, Ilkka, Leena, Kaisa, and Sanna

Contents

List of Tables   xi

List of Figures   xiii

Preface   xv

Introduction   xix

1 **Likelihood-Based Statistical Inference**   1
   1.1 Statistical evidence   2
        1.1.1 Response and its statistical model   3
        1.1.2 Sample space, parameter space, and model function .   3
        1.1.3 Interest functions   5
   1.2 Statistical inference   8
        1.2.1 Evidential statements   9
        1.2.2 Uncertainties of statements   9
   1.3 Likelihood concepts and law of likelihood   10
        1.3.1 Likelihood, score, and observed information functions   10
        1.3.2 Law of likelihood and relative likelihood function . . .   15
   1.4 Likelihood-based methods   17
        1.4.1 Likelihood region   19
        1.4.2 Uncertainty of likelihood region   20
   1.5 Profile likelihood-based confidence intervals   22
        1.5.1 Profile likelihood function   23
        1.5.2 Profile likelihood region and its uncertainty   26
        1.5.3 Profile likelihood-based confidence interval   28
        1.5.4 Calculation of profile likelihood-based confidence inter-
               vals   31
        1.5.5 Comparison with the delta method   34
   1.6 Likelihood ratio tests   37
        1.6.1 Model restricted by hypothesis   38
        1.6.2 Likelihood of the restricted model   39
        1.6.3 General likelihood ratio test statistic (LRT statistic) .   41
        1.6.4 Likelihood ratio test and its observed significance level   42
   1.7 Maximum likelihood estimate   45
        1.7.1 Maximum likelihood estimate (MLE)   45
        1.7.2 Asymptotic distribution of MLE   47

1.8 Model selection 47
1.9 Bibliographic notes 49

2 Generalized Regression Model   **51**
2.1 Examples of regression data 51
2.2 Definition of generalized regression model 69
    2.2.1 Response 70
    2.2.2 Distributions of the components of response 70
    2.2.3 Regression function and regression parameter 70
    2.2.4 Regressors and model matrix (matrices) 71
    2.2.5 Example 72
2.3 Special cases of GRM 73
    2.3.1 Assumptions on parts of GRM 73
    2.3.2 Various special GRMs 74
2.4 Likelihood inference 75
2.5 MLE with iterative reweighted least squares 76
2.6 Model checking 78
2.7 Bibliographic notes 79

3 General Linear Model   **81**
3.1 Definition of the general linear model 81
3.2 Estimate of regression coefficients 87
    3.2.1 Least squares estimate (LSE) 87
    3.2.2 Maximum likelihood estimate (MLE) 90
3.3 Test of linear hypotheses 92
3.4 Confidence regions and intervals 95
    3.4.1 Joint confidence regions for finite sets of linear combinations 95
    3.4.2 Separate confidence intervals for linear combinations . 97
3.5 Model checking 100
3.6 Bibliographic notes 103

4 Nonlinear Regression Model   **107**
4.1 Definition of nonlinear regression model 107
4.2 Estimate of regression parameters 109
    4.2.1 Least squares estimate (LSE) of regression parameters 109
    4.2.2 Maximum likelihood estimate (MLE) 112
4.3 Approximate distribution of LRT statistic 114
4.4 Profile likelihood-based confidence region 115
4.5 Profile likelihood-based confidence interval 115
4.6 LRT for a hypothesis on finite set of functions 121
4.7 Model checking 123
4.8 Bibliographic notes 124

5   Generalized Linear Model   127

   5.1   Definition of generalized linear model   127

      5.1.1   Distribution, linear predictor, and link function   127

      5.1.2   Examples of distributions generating generalized linear models   129

   5.2   MLE of regression coefficients   136

      5.2.1   MLE   136

      5.2.2   Newton-Raphson and Fisher-scoring   138

   5.3   Bibliographic notes   140

6   Binomial and Logistic Regression Model   141

   6.1   Data   141

   6.2   Binomial distribution   144

   6.3   Link functions   146

      6.3.1   Unparametrized link functions   146

      6.3.2   Parametrized link functions   149

   6.4   Likelihood inference   151

      6.4.1   Likelihood function of binomial data   151

      6.4.2   Estimates of parameters   152

      6.4.3   Likelihood ratio statistic or deviance function   154

      6.4.4   Distribution of deviance   154

      6.4.5   Model checking   156

   6.5   Logistic regression model   157

   6.6   Models with other link functions   163

   6.7   Nonlinear binomial regression model   165

   6.8   Bibliographic notes   168

7   Poisson Regression Model   169

   7.1   Data   169

   7.2   Poisson distribution   170

   7.3   Link functions   172

      7.3.1   Unparametrized link functions   172

      7.3.2   Parametrized link functions   175

   7.4   Likelihood inference   176

      7.4.1   Likelihood function of Poisson data   176

      7.4.2   Estimates of parameters   177

      7.4.3   Likelihood ratio statistic or deviance function   179

      7.4.4   Distribution of deviance   179

      7.4.5   Model checking   181

   7.5   Log-linear model   182

   7.6   Bibliographic notes   187

8 Multinomial Regression Model   **189**
  8.1   Data   189
  8.2   Multinomial distribution   191
  8.3   Likelihood function   191
  8.4   Logistic multinomial regression model   193
  8.5   Proportional odds regression model   195
  8.6   Bibliographic notes   199

9 Other Generalized Linear Regressions Models   **201**
  9.1   Negative binomial regression model   201
     9.1.1   Data   201
     9.1.2   Negative binomial distribution   203
     9.1.3   Likelihood inference   204
     9.1.4   Negative binomial logistic regression model   208
  9.2   Gamma regression model   211
     9.2.1   Data   211
     9.2.2   Gamma distribution   211
     9.2.3   Link function   213
     9.2.4   Likelihood inference   214
     9.2.5   Model checking   221

10 Other Generalized Regression Models   **225**
  10.1 Weighted general linear model   225
     10.1.1 Model   225
     10.1.2 Weighted linear regression model as GRM   226
  10.2 Weighted nonlinear regression model   229
     10.2.1 Model   229
     10.2.2 Weighted nonlinear regression model as GRM   230
  10.3 Quality design or Taguchi model   231
  10.4 Lifetime regression model   237
  10.5 Cox regression model   240
  10.6 Bibliographic notes   248

A Datasets   **251**

B Notation Used for Statistical Models   **271**

  References   **277**

  Data Index   **283**

  Author Index   **285**

  Subject Index   **287**

List of Tables

1.1 Respiratory disorders data. 3
1.2 Fuel consumption data. 6
1.3 Leukemia data. 12
1.4 Stressful events data. 17
1.5 Respiratory disorders data: Estimates and approximate 0.95-level confidence intervals of interest functions. 29
1.6 Leukemia data: Estimates and approximate 0.95-level confidence intervals of interest functions. 31
1.7 Bacterial concentrations data. 38
1.8 Leukemia data: Estimates and approximate 0.95-level confidence intervals of parameters. 45

2.1 Leaf springs data. 53
2.2 Sulfisoxazole data. 57
2.3 Clotting time data. 59
2.4 Remission of cancer data. 62
2.5 Smoking data. 64
2.6 Coal miners data. 67

3.1 Tensile strength data. 83
3.2 Red clover plants data: Estimates and exact 0.95-level confidence intervals (t-intervals) for pairwise differences of means. 100
3.3 Waste data: Estimates and exact 0.95-level confidence intervals (t-intervals) of interaction contrasts. 101

4.1 Puromycin data. 108
4.2 Puromycin data: Estimates and approximate 0.95-level confidence intervals of interest functions. 117
4.3 Puromycin data: Estimates and approximate 0.95-level confidence intervals based on $F[1,10]$distribution. 117
4.4 Sulfisoxazole data: Estimates and approximate 0.95-level confidence intervals of interest functions. 120
4.5 Sulfisoxazole data: Estimates and approximate 0.95-level confidence intervals based on $F[1,8]$distribution. 120

5.1 NormalModel. 130
5.2 PoissonModel. 130

5.3   BinomialModel.   132
5.4   GammaModel.   134
5.5   InverseGaussianModel.   134

6.1   O-ring data: Estimates and approximate 0.95-level confidence
        intervals of model parameters.   159
6.2   Insulin data   161
6.3   Insulin data: Estimates and approximate 0.95-level confidence
        intervals of model parameters.   162
6.4   Remission of cancer data: Estimates and approximate 0.95-
        level confidence intervals of model parameters.   165
6.5   Malaria data   167

7.1   Lizard data: Estimates and approximate 0.95-level confidence
        intervals of model parameters for the model with two factor
        interactions.   185
7.2   Lizard data: Estimates and approximate 0.95-level confidence
        intervals of model parameters for the model with selected two
        factor interactions.   187

8.1   Smoking data: Estimates and approximate 0.95-level confi-
        dence intervals of model parameters.   194
8.2   Smoking data: Estimates and approximate 0.95-level confi-
        dence intervals of interest functions in the submodel.   195
8.3   Coal miners data: Estimates and approximate 0.95-level con-
        fidence intervals of interest functions.   199

9.1   Pregnancy data: Estimates and approximate 0.95-level confi-
        dence intervals of interest functions.   209
9.2   Brain weights data: Estimates and approximate 0.95-level
        confidence intervals of model parameters.   219
9.3   Brain weights data: Estimates and approximate 0.95-level
        confidence intervals of interest functions.   219

10.1   Fuel consumption data: Estimates and approximate 0.95-level
        confidence intervals of model parameters.   227
10.2   Puromycin data: Estimates and approximate 0.95-level confi-
        dence intervals of interest functions in the treated group. . .   231
10.3   Leaf springs data: Levels of factors.   234
10.4   Leaf springs data: Model matrix for means.   234
10.5   Leaf springs data: Model matrix for standard deviations. . .   235
10.6   Leaf springs data: Estimates and approximate 0.95-level con-
        fidence intervals of model parameters.   236
10.7   Leaf springs data: Estimates and approximate 0.95-level con-
        fidence intervals of model parameters in submodel.   236

10.8   Leaf springs data: Estimates and approximate 0.95-level confidence intervals of standard deviations for combinations of heating time and furnace temperature factors.   236

10.9   Leaf springs data: Estimates and approximate 0.95-level confidence intervals of means for combinations of oil temperature, transfer time, and furnace temperature factors.   237

10.10 Carcinogen data.   238

10.11 Carcinogen data: Estimates and approximate 0.95-level confidence intervals of model parameters.   239

10.12 Carcinogen data: Estimates and approximate 0.95-level confidence intervals of interest functions in the submodel.   240

10.13 Gastric cancer data.   243

10.14 Gastric cancer data: Estimates and approximate 0.95-level confidence intervals of interest functions.   245

10.15 Leukemia data: Estimates and approximate 0.95-level confidence intervals of interest functions.   247

10.16 Acute myelogeneous leukemia data: Estimates and approximate 0.95-level confidence intervals of model parameters. . .   249

A.1   Acute myelogeneous leukemia data.   252
A.2   Bacterial concentrations data.   253
A.3   Brain weights data.   254
A.4   Species names.   255
A.5   Carcinogen data.   255
A.6   Clotting time data.   256
A.7   Coal miners data.   256
A.8   Fuel consumption data.   257
A.9   Gastric cancer data.   257
A.10   Insulin data.   258
A.11   Leaf springs data.   259
A.12   Leukemia data.   259
A.13   Lizard data.   260
A.14   Malaria data.   261
A.15   O-ring data.   262
A.16   Pregnancy data.   263
A.17   Puromycin data.   264
A.18   Red clover plants data.   265
A.19   Remission of cancer data.   266
A.20   Respiratory disorders data.   266
A.21   Smoking data.   267
A.22   Stressful events data.   268
A.23   Sulfisoxazole data.   269
A.24   Tensile strength data.   269
A.25   Waste data.   270

B.1   Discrete distributions.   272
B.2   Continuous distributions.   273
B.3   Basic statistical models.   274
B.4   Regression models.   275

List of Figures

1.1   Control group of respiratory disorders data: Point probability function   6

1.2   Fuel consumption data: Scatter plots   7

1.3   Fuel consumption data: 3-way scatter plot   7

1.4   Leukemia data: Scatter plot   12

1.5   Leukemia data: Contour plot of the log-likelihood function.   14

1.6   Leukemia data: Contour plot of score   15

1.7   Stressful events data: Scatter plots   18

1.8   Stressful events data: Plots of likelihood   18

1.9   Respiratory disorders data: Contour plot   20

1.10   Fuel consumption data: Plot of uncertainty   22

1.11   Fuel consumption data: Plot of likelihood for given mean . .   24

1.12   Fuel consumption data: Plots of relative likelihood and log-likelihood functions.   26

1.13   Fuel consumption data: Plot of likelihood   27

1.14   Respiratory disorders data: Plots of log-likelihoods of interest functions.   30

1.15   Leukemia data: Plots of log-likelihoods of interest functions   32

1.16   Respiratory disorders data: Iteration steps for interval . . .   34

1.17   Leukemia data: Iteration steps of interval for AUC under ROC.   35

1.18   Fuel consumption data: Plot of log-likelihood.   37

1.19   Bacterial concentrations data: Plot of data   39

1.20   Respiratory disorders data: Contour plot.   46

1.21   Respiratory disorders data: Iteration.   47

1.22   Leukemia data: Contour plot.   48

1.23   Leukemia data: Iteration steps.   48

2.1   Leaf springs data: Means   54

2.2   Leaf springs data: Standard deviations   55

2.3   Sulfisoxazole data: Scatter plot   58

2.4   Clotting time data: Scatter plot   60

2.5   Remission of cancer data: Scatter plot of logits   62

2.6   Smoking data: Plot of logits   65

2.7   Coal miners data: Plot of cumulative logits   67

3.1 Tensile strength data: Scatter plot 84
3.2 Fuel consumption data: Regression plane 85
3.3 Fuel consumption data: Regression surface 86
3.4 Fuel consumption data: Confidence region. 97
3.5 Fuel consumption data: Plot of log-likelihood. 98
3.6 Red clover plants data: Scatter plot 99
3.7 Waste data: Scatter plot 101
3.8 Tensile strength data: Residuals 103
3.9 Fuel consumption data: Residuals 105

4.1 Puromycin data: Scatter plot 109
4.2 Puromycin data: Confidence region 116
4.3 Puromycin data: Plots of log-likelihoods of interest functions. 118
4.4 Sulfisoxazole data: Plots of log-likelihoods of interest func-
     tions. 122
4.5 Puromycin data: Residuals 125
4.6 Sulfisoxazole data: Residuals 126

5.1 NormalModel: Densities 131
5.2 PoissonModel: Point probability functions 132
5.3 BinomialModel: Point probability functions 133
5.4 GammaModel: Densities 135
5.5 InverseGaussianModel: Densities 137

6.1 O-ring data: Scatter plots 143
6.2 BinomialModel: Point probability function 145
6.3 Binomial link functions. 147
6.4 Inverse of cumulative distribution function of CauchyModel 148
6.5 Parametrized binomial link function. 150
6.6 Empirical distribution functions. 155
6.7 O-ring data: QQ-plots of residuals. 158
6.8 O-ring data: Scatter plots of modified residuals. 158
6.9 O-ring data: Plots of log-likelihoods of interest functions. . . 160
6.10 Insulin data: Scatter plot 161
6.11 Insulin data: Plots of residuals. 163
6.12 Insulin data: Plots of log-likelihoods of interest functions. . . 164
6.13 Remission of cancer data: Residual plots. 166
6.14 Malaria data: Scatter plot 167

7.1 Lizard data: Scatter plots. 171
7.2 PoissonModel: Point probability function 173
7.3 Comparison of link functions for Poisson distribution. 174
7.4 Empirical distribution functions. 181
7.5 Stressful events data: Residuals. 183
7.6 Lizard data: Plots of residuals. 186

8.1   Smoking data: Plots of log-likelihoods of interest functions. .   196
8.2   Coal miners data: Plots of log-likelihoods of interest functions. 200

9.1   Pregnancy data: Observed point probability functions . . .   202
9.2   Point probability functions of negative binomial distributions   205
9.3   Pregnancy data: Contour plot of the log-likelihood function   207
9.4   Pregnancy data: Plots of log-likelihoods of interest functions.   210
9.5   Brain weights data: Plot of data   212
9.6   Densities of gamma distribution   213
9.7   Plots of link functions for gamma distribution   215
9.8   Brain weights data: Plots of log-likelihood of model parame-
       ters.   218
9.9   Brain weights data: Plots of log-likelihoods of interest func-
       tions.   220
9.10   Brain weights data: Plots of residuals.   223
9.11   Brain weights data: Plots of modified residuals and mean
       brain weight.   224

10.1   Fuel consumption data: Plots of log-likelihoods of model pa-
       rameters.   228
10.2   Puromycin data: Plots of log-likelihoods of interest functions
       in the treated group.   232
10.3   Carcinogen data: Plot of data.   238
10.4   Carcinogen data: Plots of log-likelihoods of interest functions
       in the submodel.   241
10.5   Gastric cancer data: Plot of data.   244
10.6   Gastric cancer data: Plots of log-likelihoods of interest func-
       tions.   246
10.7   Leukemia data: Plots of log-likelihoods of interest functions.   248
10.8   Acute myelogeneous leukemia data: Scatter plot of data . .   249
10.9   Acute myelogeneous leukemia data: Plots of log-likelihoods of
       model parameters.   250

Symbol Description

$A \times B$	Cartesian product of sets A and B.	ω_i	i^{th} component of parameter vector.
d	Dimension of parameter space.	$\Pr(A; \omega)$	Probability of event A.
ϕ	Density function of standard normal distribution.	ψ	Interest function.
		$p(y; \omega)$	Model function.
Φ	Cumulative distribution function of standard normal distribution.	$\mathbb{R}^n$	n-dimensional Euclidean space.
$\ell(\omega; y)$	Log-likelihood function based on observation y.	$r(\omega; y)$	Relative log-likelihood function based on observation y.
$L_{(}\omega; y)$	Likelihood function based on observation y.	$R(\omega; y)$	Relative likelihood function based on observation y.
$\mathcal{M}$	Statistical model.	σ	Standard deviation.
μ	Mean.	$\mathcal{Y}$	Sample space.
n	Dimension of sample space.	y	Response vector.
Ω	Parameter space.	y_{obs}	Observed response vector.
ω	Parameter vector.	y_i	i^{th} component of response vector.
$\hat{\omega}$	Maximum likelihood estimate of parameter vector.		

Preface

During the years 1980 through 1990, I did research on simultaneous confidence intervals in linear models. Towards the end of this period, I started to think about how simultaneous confidence interval procedures could be generalized to general statistical models, especially generalized linear models. It soon became apparent that the "right" way to generalize simultaneous confidence intervals is to use the likelihood approach. Similar development had already appeared in nonlinear regression models when Bates and Watts (1988) generalized t-intervals to "t"-intervals. These intervals are a special case of the so-called profile likelihood-based confidence intervals. In addition to what Bates and Watts had considered, the need to consider nonlinear real-valued functions of components of the parameter vector arose as well.

The standard algorithm for the construction of profile likelihood-based confidence intervals is based on the determination of the profile likelihood function and its derivative. Clarke (1987) presented for general interest function in the nonlinear regression model and Venzon and Moolgavkar (1988) for a component of the parameter vector in the general statistical model a different algorithm, which involved only the likelihood and score functions of the statistical model and interest function and its gradient. Around 1993, I realized that their algorithm was applicable in any parametric statistical model for any smooth enough interest function, that is, any interest function, which is continuous with a nonvanishing gradient. This algorithm transforms the determination of the confidence interval into finding minimum and maximum of the interest function on the boundary of a certain likelihood-based confidence region for the whole parameter vector, that is, into a problem of the constrained optimization of a generally nonlinear real-valued function, interest function, with one nonlinear real-valued function, likelihood function, giving the constraint. First I implemented practical calculations using a well-known statistical package, but because for various reasons, including using adjusted profile likelihood functions, I needed a platform powerful in symbolic calculation, especially symbolic differentiation, I moved to using *Mathematica* sometime before 1996. I will explain more about the implementation at the end of this preface.

In the course of implementation of profile likelihood-based confidence interval calculations, keeping in mind that it is applicable in the case of general statistical models and general smooth interest functions, I decided to build the calculation system so that it would make it easy to consider very general statistical models. This meant especially considering the possibility to define parametric generalized/general regression models, which Lindsey (1974a, 1974b, 1996, 1997) had already been studying. In addition to Lindsey, the concept of the generalized regression model has been discussed by Davison (2003) in the case of one-parameter generalized regression models. Thus, the implementation of regression models has been done with Lindsey's ideas in mind.

This book is the product of these two developments, namely profile likelihood-based confidence intervals and generalized regression models. It contains material that I have been presenting in courses during the last ten years, especially the likelihood framework with its justification. During the Fall of 2006, I gave a course on generalized linear models, but the material in that course was more general and thus the basis of this book.

This book is intended for senior undergraduate and graduate students, but material on confidence intervals should be useful to applied statisticians as well. The readers should be familiar with common statistical procedures such as t-intervals, simple linear regression, and analysis of variance. The reader is also assumed to have basic knowledge of calculus and linear algebra and to have taken an introductory course in probability and theoretical statistics.

I would like to acknowledge the assistance of the editors at Taylor and Francis in completing this project. I especially wish to thank David Grubbs and Michele Dimont for their continuing support.

I thank my family, my children Silja, Lenni, and Ilkka, and my wife Eliisa, for the patience they have had during the years of writing the *SIP* package and especially during the last year when I was writing the book.

Software

When implementing the calculation of profile likelihood-based confidence intervals I started to use the *Mathematica* functions I had developed in my courses little by little, especially in courses on statistical inference and generalized linear models. Some years went by and I noticed that I had enough material to start to develop a *Mathematica* package, *Statistical Inference Package SIP* (Uusipaikka, 2006). I had been using various statistical packages in my teaching from around 1975, but had always been annoyed by the fact that the notation I used in the class when discussing the theory of statistical inference was totally different from what I had to use when I discussed and used the statistical procedures of statistical packages. So from the beginning I decided to build a system for doing statistical inference, where the notation would be as much as possible that used in books and lectures.

The basic structure of *SIP* was constructed so that it would be easy and intuitively straightforward to build more complicated and as general as possible statistical models from simpler statistical models and distributions. For example, in *SIP* it is possible to analyze a sample from any distribution (30 in *SIP*) or model (in principle unlimited number) defined in the system by defining a sampling model. *SIP* was released February 8, 2006. Additional information on *SIP* can be obtained from the Wolfram Research, Inc. Website

http://www.wolfram.com/products/applications/sip/.

The third aim of the book is to explain the usage of *SIP* in analysis of statistical evidence consisting of data and its regression model. The main text, however, contains only numerical and graphical results obtained by *SIP*, and *Mathematica* code used to get those results is available from the Website

http://users.utu.fi/esauusi/kirjat/books/CIGRM/.

The Website also includes *R* code for selected examples from the main text. In many examples, however, because of certain restrictions of *R*, the exact same analysis is out of reach. Also, because of the totally different philosophy of this package when compared to *SIP*, no instructions on the usage of it are provided. There are good books for that purpose, such as Dalgaard (2002) and Faraway (2005, 2006).

A DVD containing restricted versions of *Mathematica* and *SIP* is distributed with the book. Using them with downloaded files, readers can do the calculations themselves. *SIP* is an add-on package of *Mathematica*®, a trademark of Wolfram Research, Inc. The *SIP* package has been developed mainly to assist applied statisticians to make inferences on real valued parameter functions of interest. The package contains procedures for the calculation of profile likelihood-based confidence intervals and their uncertainties for the unknown value of a real-valued interest function. It also includes functions to make likelihood ratio tests concerning some given hypothesis about the parameters of the statistical model. The *SIP* package can handle any linear or smooth nonlinear interest function of the parameters in the case of interval estimation or hypothesis testing.

The *SIP* package contains a large collection of regression models:

- LinearRegressionModel

- LogisticRegressionModel

- PoissonRegressionModel

- GeneralizedLinearRegressionModel

- MultivariateLinearRegressionModel

- MultinomialRegressionModel

- NonlinearRegressionModel

- RegressionModel

In the first six of these, a parameter (or parameters), usually means (or mean), of the components of the response is assumed to depend linearly or through some monotone transformation on regression parameters. The models differ with respect to the form of the response distribution. The models `NonlinearRegressionModel` and `RegressionModel` allow the dependence between regression parameters and the parameters of the response model to be properly nonlinear. In `NonlinearRegressionModel` the distributions of the components of the response are assumed to be normal. `RegressionModel` is a very general function for defining regression models where the only assumptions are the independence of the components of the response and the dependence of their distributions on the common parameters. Thus, `RegressionModel` implements the generalized regression models.

Esa Uusipaikka
Raisio, Finland

Introduction

Regression model is undoubtedly the most often used type of statistical model. The reason for this is that in empirical research the perennial question is how a given characteristic or set of characteristics of an item under study is related or affects another characteristic or set of characteristics of that same item. The former characteristics are usually called *explaining variables* or *factors* and the latter *explained variables* or *responses*. Application of regression models consists of assuming that there exists some functional relation between factors and responses, but because responses almost invariably contain uncontrolled haphazard variation, this relation must be expressed as a relation between the factors and a parameter or parameters of the statistical model of the response. The function relating the values of the factors to the values of parameters is called *regression function*.

There is a huge number of different types of regression models from very simple to extremely complex ones. In this book among others, the following regression models will be discussed:

- *general linear model* (GLM)

- *nonlinear regression model*

- *generalized linear regression model* (GLIM)

- *logistic regression model*

- *Poisson regression model*

- *multinomial regression model*

- *negative binomial regression model*

- *gamma regression model*

- *lifetime regression model*

One of the aims is to introduce a unified representation of various types of regression models. This unified representation is called *generalized (parametric) regression model* (GRM) and it tries to capture the main characteristics of regression models, that is,

1. (conditional) independence of the components of the response vector (given the values of factors) and

2. dependence of a component or components of the parameter vector for the response on common finite dimensional *regression parameter* through the regression function also involving the values of factors.

A unified treatment of regression models would be of little value if there did not exist some unified way of producing statistical inferences for data modeled by the generalized regression models. Thus, the second aim is to use a unified approach for doing statistical inference from statistical evidence consisting of data and its statistical model. The unified approach is the *likelihood-based approach*, which is applicable in all parametric statistical models including generalized parametric regression models.

Significance tests and in far less proportion *Wald-type confidence intervals* for the regression parameters have thus far been the main methods of statistical inference in the case of regression analysis. In addition, the so-called *delta method* is sometimes used to get confidence intervals for functions of regression parameters. It is well known that Wald-type intervals including those obtained with the delta method have many defects. Because of this, in this book the approach to statistical inference consists mainly of calculating *profile likelihood-based confidence intervals* for parameter functions of interest and in a lesser quantity performing *likelihood ratio tests* of hypotheses on regression parameters, the latter of which has been around for over 50 years but the former in its very general form considered in this book is new. Both of these inference procedures also involve *maximum likelihood estimation* of the model parameter. Likelihood-based inference generally involves approximations using asymptotics of some kind. This is understandable because of its general applicability. Likelihood-based confidence intervals and likelihood ratio tests, however, are (after small modifications) equivalent to well-known exact intervals and tests in the most important special cases, for example, in GLM.

As stated above, the emphasis is on making statistical inferences in the form of profile likelihood-based confidence intervals and less attention is given to model selection and criticism as these latter two are well covered in numerous books on regression analysis, such as Searle (1971), Seber (1977), Draper and Smith (1981), Weisberg (1985), Wetherill (1986), Sen and Srivastava (1990), and Jørgensen (1997a). Thus, for pedagogical reasons sometimes models that may not be the best fitting ones are used. It is readily admitted, however, that inferences based on badly fitting models may be totally inappropriate.

In Chapter 1 likelihood-based statistical inference is introduced. Starting from statistical evidence consisting of the actual value of the response and its statistical model, first likelihood concepts are reviewed and then the law of likelihood is introduced. Based on the law of likelihood, methods of constructing confidence regions, profile likelihood-based confidence intervals, and likelihood ratio tests are considered. The latter two are applicable to any linear or smooth nonlinear function of the parameter vector. Then maximum likelihood estimation is covered, not because it would be a proper inference procedure, but because it is an integral part of confidence interval and likelihood ratio test construction. Also, model selection through Akaike's Information Criterion is presented. This rather long chapter is included more as a reference, but because the likelihood-based approach to statistical inference is not common knowledge, instead of being given as an appendix, this form was chosen to emphasize its importance to the rest of the book.

Chapter 2 starts with an introduction of the generalized regression model (GRM), which informally can be described as a statistical model for statistically independent observations such that models for these independent observations depend on a common parameter vector. After that, special cases of GRMs, including all the familiar regression models, are characterized as GRMs. The chapter ends with a presentation of iterated weighted least squares algorithm for the maximum likelihood estimation of the regression parameters.

From Chapter 3 on various special cases of GRMs are discussed starting with the GLM in Chapter 3. Then in Chapter 4 the nonlinear regression model is covered.

Chapter 5 introduces GLIM, which is a generalization of GLM for some nonnormal distributions for the response. Then from Chapter 6 on various special cases of GLIMs are considered. First, in Chapter 6, logistic regression, where the distributions of the components of the response are binomial distributions, is considered, and then in Chapter 7 Poisson distribution provides the model for the components of the response vector. Chapter 8 covers the case of multinomially distributed components of the response vector.

Chapter 9 introduces two less frequently used distributions, namely negative binomial and gamma distributions. Finally, Chapter 10 contains various special cases including weighted general linear and nonlinear models, a model for quality design data, lifetime regression models, and the proportional hazard regression model.

Chapter 1

Likelihood-Based Statistical Inference

> Statistics concerns what can be learned from data. Applied statistics comprises a body of methods for data collection and analysis across the whole range of science, and in areas such as engineering, medicine, business, and law — wherever variable data must be summarized, or used to test or confirm theories, or to inform decisions (Davison 2003, p. 1).

Statistical inference methods and, more broadly, statistical analysis methods form a huge collection of formal analytical and graphical tools for dealing with empirical data. In a broad sense all of these methods have been constructed for the purpose of extracting information from data. There exists, however, certain differences between them related to the situations where they are used. There are at least three main classes of methods, namely,

1. methods used as tools for making decisions, the results of which are actions in the real world,

2. methods used to update beliefs about some phenomenon based on observations on that same phenomenon, and

3. methods to find out what explanations or theories about the phenomenon, which is under investigation, are supported by the empirical data.

Clearly, the latter two must ultimately have consequences in some form of action in the real world, otherwise they would not have any value. Results of the second and third class of tools, however, do not contain actual decisions; instead, for the time being, one is satisfied with updated beliefs or knowledge about support, respectively. Also, because results of the third class of tools are always evaluated by people, it is inevitable that knowledge of support will result in updated beliefs for those using them, but that updating is not included in the tools.

The appropriate tool to be used clearly depends on the situation where it is used. Decision-making tools are needed in situations where it is necessary to produce some real action. Tools for updating beliefs are needed in situations where opinions matter. Finally, tools for finding out how much different explanations or theories are supported by empirical data are needed in situations where results of studies must be told to a larger audience and updating beliefs is left to each individual using the results of the analysis. In this book, the tools are presented in the framework of assessing the support given to various explanations by the empirical data and its statistical model. The tools appearing in this book may be valuable in the other two situations as well, but their application in those situations will not be discussed.

As Pawitan (2001, p. v) notes, statistical tools contain methods for

1. *planning* studies and experiments,

2. *describing* numerically and graphically empirical data,

3. *modeling* variability and relations of studied quantities,

4. *inference* to assess the amount and uncertainty of interesting associations and effects, and

5. *model selection and criticism* to assess the sensibility of the used model.

In this book, main emphasis is on modeling and inference with some discussion also on model selection and criticism.

The vast majority of all statistical research is concerned with the construction of suitable statistical distributions and models to be used by researchers in all scientific areas to analyze their empirical data. This research will never be complete because the studies of scientists become more involved as their understanding of their fields increases. Among statistical models, regression models have been historically and quantitatively the most important ones. They will be discussed thoroughly starting in the next chapter.

In Section 1.1 the concept of statistical evidence consisting of empirical data and its statistical model is discussed. In Section 1.2 evidential statements and their uncertainties are introduced as components of statistical inference. Then in Section 1.3 likelihood concepts and law of likelihood are discussed. Section 1.4 starts the presentation of likelihood-based inference methods by introducing the likelihood-based confidence region. Then in Section 1.5 likelihood-based confidence intervals are introduced with a new method for the calculation of these intervals in very general situations. Sections 1.6 and 1.7 contain discussions of likelihood ratio test and maximum likelihood estimation. Finally, in Section 1.8 a brief introduction to model selection is presented.

1.1 Statistical evidence

The key feature of a statistical model is that variability is represented using probability distributions, which form the building-blocks from which the model is constructed. Typically it must

TABLE 1.1: Respiratory disorders data.

Treatment	Favorable	Unfavorable
Placebo	16	48
Test	40	20

accommodate both random and systematic variation. The randomness inherent in the probability distribution accounts for apparently haphazard scatter in the data, and the systematic pattern is supposed to be generated by structure in the model (Davison 2003, p. 1).

The above quotation from Anthony Davisons book *Statistical Models* (2003) is an apt characterization of the role of probability in statistical analysis.

1.1.1 Response and its statistical model

Statistical inference is based on *statistical evidence*, which has at least two components. The two necessary components consist of *observed data* and its *statistical model*.

Example 1.1 Test group of respiratory disorders data

Table 1.1 summarizes the information from a randomized clinical trial that compared two treatments (test, placebo) for a respiratory disorder (Stokes *et al.* 1995, p. 4). The columns in the table contain values of treatment, count of favorable cases, and count of unfavorable cases.

Let us consider as an example the number of respiratory disorders in the test group of the respiratory disorders dataset; that is, let the observed data consist of the count (40) of favorable cases among the total number (60) of patients in the test group. That part of the data that contains random variation is often called *response*. One possible statistical model for the count of favorable cases is obtained by considering it as the number of "successes" in 60 statistically independent "trials" with common unknown "success" probability ω, that is, assuming that it is an observation from some binomial distribution with 60 trials. ⬛

1.1.2 Sample space, parameter space, and model function

Because by definition the response part of the data contains random variation, the actual observed value of the response is one among a set of plausible values for the response. The set of all possible values for the response is called the *sample space* and denoted by $\mathcal{Y}$. The generic point y in the sample space

is called the *response vector*. Sometimes the response has a more complicated structure, but even in these cases the somewhat imprecise expression
"response vector" is used. Because in likelihood-based statistical inference
the structure of the sample space has minor effect, this practice should not
be confusing.

The random variation contained in the response is described by either a
point probability function or density function defining a probability distribution in the sample space $\mathcal{Y}$. Because the purpose of statistical inference is to
give information on the unknown features of the phenomenon under study,
these unknown features imply that the probability distribution of the response
is not completely known. Thus, it is assumed that there is a set of probability distributions capturing the important features of the phenomenon, among
which features of primary interest are crucial, but also so-called secondary
ones are important. It is always possible to index the set of possible probability distributions and in statistical inference this index is called *parameter*.
The set of all possible values for the parameter is called the *parameter space*
and denoted by Ω. The generic point ω in the parameter space is called the
parameter vector.

In this book, only so-called *parametric statistical models* are considered. In
parametric statistical models the value of the parameter is determined by the
values of finitely many real numbers, which form the finite dimensional parameter vector. Thus, the parameter space is a subset of some finite dimensional
space $\mathbb{R}^d$ where d is a positive integer denoting the dimension of the parameter space. Sometimes the parameter has a more complicated structure, but in
the following it is always assumed that there is a one-to-one correspondence
between the parameter and parameter vector. However, in all cases parameter
space means the set of all possible values of the parameter vector. Symbols
used to index probability distributions of statistical models sometimes refer
to the parameter and other times to the parameter vector. The one-to-one
relation between a parameter and its corresponding parameter vector implies
that no confusion should arise with this practice and the meaning of symbols
is, in most cases, clear because of the context in which they appear.

There is a probability distribution defined for every value of the parameter
vector ω. Thus, a statistical model is defined by a *model function* $p(y; \omega)$,
which is a function of the observation vector y and the parameter vector ω
such that for every fixed value of parameter vector the function defines a
distribution in the sample space $\mathcal{Y}$. The function of the parameter vector
obtained by fixing the observation vector (equal to actual response) is called
the (observed) likelihood function and is very important in the likelihood-
based statistical inference. The statistical model consists of the sample space
$\mathcal{Y}$, the parameter space Ω, and the model function $p(y; \omega)$. It is denoted by

$$\mathcal{M} = \{p(y; \omega) : y \in \mathcal{Y}, \omega \in \Omega\} \tag{1.1}$$

or by the following mathematically more correct notation

$$\mathcal{M} = \{p(\cdot; \omega) : \omega \in \Omega\}, \tag{1.2}$$

where $p(\cdot; \omega)$ denotes the point probability or density function on the sample space $\mathcal{Y}$ defined by the parameter vector ω. Even though the former notation is slightly confusing it is mainly used.

Example 1.2 Test group of respiratory disorders data (cont'd)

The random variation in the test group of the respiratory disorder data example was assumed to follow the binomial distribution with 60 trials and the unknown "success" probability ω as the unknown parameter. The sample space $\mathcal{Y}$ consists of the set of integers $0, 1, 2, \ldots, 60$, the parameter space Ω is the open interval $(0, 1) \subset \mathbb{R}^1$, and the model function has the form

$$p(y; \omega) = \binom{60}{y} \omega^y (1 - \omega)^{60-y}$$

for all $y \in \mathcal{Y}$ and $\omega \in \Omega$. ▯

Example 1.3 Placebo group of respiratory disorders data

In the placebo group of the respiratory disorders data example the sample space $\mathcal{Y}$ consists of the set of integers $0, 1, 2, \ldots, 64$, and the parameter space Ω is the open interval $(0, 1) \subset \mathbb{R}^1$. Assuming that the random variation in the number of favorable outcomes can be described by binomial distribution, the statistical model $\mathcal{M}$ is denoted by

$$\mathcal{M} = \text{BinomialModel}[64, \omega]$$

and the model function $p(y; \omega)$ has the form

$$\binom{64}{y} \omega^y (1 - \omega)^{64-y}.$$

Figure 1.1 contains a plot of the point probability function for the value 0.4 of the parameter ω of the statistical model $\mathcal{M}$, that is, for the success probability 0.4. ▯

1.1.3 Interest functions

Statistical inference concerns some characteristic or characteristics of the phenomenon from which the observations have arisen. The characteristics of the phenomenon under consideration are some functions of the parameter

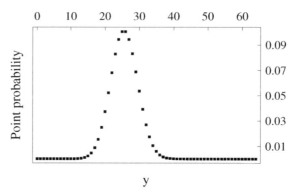

FIGURE 1.1:　Control group of respiratory disorders data: Plot of the point probability function of binomial distribution with 64 trials for the value 0.4 of success probability ω.

and are called the *parameter functions of interest*. If a subset of the components of the parameter vector form the interest functions, then the rest of the components are said to be *nuisance parameters*.

Example 1.4 **Fuel consumption data**

The data in Table 1.2 were collected to investigate how the amount of fuel oil required to heat a home depends upon the outdoor air temperature and wind velocity (Kalbfleisch 1985, p. 249). Columns of data matrix contain values of fuel consumption, air temperature, and wind velocity. Figure 1.2 and Figure 1.3 show plots of observed values of all variables against each other and a three-dimensional scatter plot, respectively.

Assume that the fuel consumption measurements may be modeled as a sample from some normal distribution with unknown mean μ and unknown

TABLE 1.2:　Fuel consumption data.

Consumption	Temperature	Velocity
14.96	-3.0	15.3
14.10	-1.8	16.4
23.76	-10.0	41.2
13.20	0.7	9.7
18.60	-5.1	19.3
16.79	-6.3	11.4
21.83	-15.5	5.9
16.25	-4.2	24.3
20.98	-8.8	14.7
16.88	-2.3	16.1

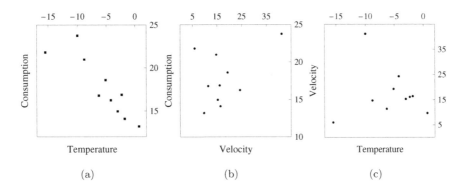

FIGURE 1.2: Fuel consumption data: Scatter plots of variables of fuel consumption data. (a) Fuel consumption against air temperature, (b) fuel consumption against wind velocity, and (c) wind velocity against air temperature.

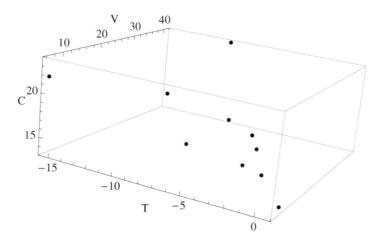

FIGURE 1.3: Fuel consumption data: Three-dimensional scatter plot of fuel consumption against air temperature and wind velocity. Symbols T, V, and C denote air temperature, wind velocity, and fuel consumption, respectively.

standard deviation σ. The appropriate statistical model $\mathcal{M}$ is denoted by

$$\mathcal{M} = \text{SamplingModel}[\text{NormalModel}[\mu, \sigma], 10]$$

with sample space $\mathcal{Y} = \mathbb{R}^{10}$ and parameter space $\Omega = \mathbb{R} \times (0, \infty) \subset \mathbb{R}^2$.

Usually in a situation like this mean μ is the parameter of interest and the standard deviation σ is a nuisance parameter. There exists, however, no theoretical reason why the standard deviation σ could not be the parameter of interest and the mean μ the nuisance parameter. In this case a more concrete and more easily interpretable quantity would, for example, be the probability that fuel consumption is greater than 20, that is, the parameter function

$$\psi = g(\mu, \sigma) = 1 - \Phi\left(\frac{20 - \mu}{\sigma}\right),$$

where Φ is the cumulative distribution function of the standard normal distribution. $\qquad\qquad\qquad\qquad\qquad\qquad\qquad\qquad\qquad\qquad\qquad$ ⬚

Example 1.5 Respiratory disorders data (cont'd)

Assume that numbers of trials have been chosen beforehand and the statistical units are independent of each other such that those in *Placebo* group have common success probability γ and those in *Test* group have possibly different common success probability λ. The statistical model for the data is denoted by

$$\mathcal{M} = \text{IndependenceModel}[\{\text{BinomialModel}[64, \gamma], \text{BinomialModel}[60, \lambda]\}]$$

with sample space $\mathcal{Y} = \{0, 1, \ldots, 64\} \times \{0, 1, \ldots, 60\}$, parameter space $\Omega = (0, 1) \times (0, 1) \subset \mathbb{R}^2$, and model function $p(y; \omega)$ of the following form

$$\binom{64}{x} \gamma^x (1 - \gamma)^{64-x} \binom{60}{z} \lambda^z (1 - \lambda)^{60-z},$$

where $y = (x, z) \in \mathcal{Y}$ and $\omega = (\gamma, \lambda) \in \Omega$.

Now the parameter function of interest might be the ratio of odds for favorable outcome in the test group to that in the placebo group, that is, the ratio $\psi = g(\gamma, \lambda) = \frac{\lambda(1-\gamma)}{(1-\lambda)\gamma}$. A more easily interpretable interest function would be the increase in the number of favorable cases among a total of say 100,000 potential cases $\psi = g(\gamma, \lambda) = 100,000(\lambda - \gamma)$. $\qquad\qquad\qquad\qquad$ ⬚

1.2   Statistical inference

A limited amount of data can be misleading, and therefore any general conclusions drawn will be subject to uncertainty. Mea-

surement of the extent of this uncertainty is an important part of
the problem (Kalbfleisch 1985, p. 1).

Given the statistical evidence consisting of observed data and its statistical
model, one has to decide what to do with that evidence, that is, how to
produce inferences on matters of interest. In addition, an essential part of
statistical inference is the assessment of uncertainties of the statements made.

1.2.1   Evidential statements

The result of a statistical inference procedure is a *collection of statements*
concerning the unknown values of the parameter functions of interest. The
statements are based on statistical evidence. The important problem of the
theory of statistical inference is to characterize the form of statements that a
given statistical evidence supports.

Example 1.6 Respiratory disorders data (cont'd)

For example, in the respiratory disorders data the statistical evidence includes counts of 16 and 40 favorable outcomes in 64 and 60 trials in case of
Placebo and *Test* groups, respectively.

The second part of the statistical evidence consists of the following statistical model

$$\mathcal{M} = \text{IndependenceModel}[\{\text{BinomialModel}[64, \gamma], \text{BinomialModel}[60, \lambda]\}].$$

A parameter function of interest might be the risk ratio, that is, the ratio
$\psi = g(\gamma, \lambda) = \frac{\lambda}{\gamma}$ of success probabilities. The fundamental problem in this
example is to derive statements concerning the risk ratio that the given sta-
tistical evidence supports and to assess the uncertainties of those supported
statements.

The evidential statements based on the statistical evidence might consist
of a single statement "the risk ratio $\frac{\lambda}{\gamma}$ is equal to 1" or of all statements "the
risk ratio $\frac{\lambda}{\gamma}$ is between $\frac{8}{3} - \delta$ and $\frac{8}{3} + \delta$" for some positive real number δ.   ⬚

1.2.2   Uncertainties of statements

The *uncertainties* of the statements are also an essential part of the inference. The statements and their uncertainties are both based on statistical
evidence and together form the *evidential meaning* of the statistical evidence.
The theory of statistical inference has to characterize the way by which the
uncertainties of the supported statements are determined. In the following we
consider one approach to solve these problems, namely, the *likelihood approach
to statistical inference* or the *likelihood method*.

1.3 Likelihood concepts and law of likelihood

> The likelihood supplies a natural order of preference among the possibilities under consideration (Fisher 1956, p. 70).

In this book tools and theory of so-called likelihood-based statistical inference are discussed more thoroughly to provide a solid foundation to the developments considered in the book. There are two reasons for this. First, by covering some developments in likelihood-based inference more versatile inference tools become available. Second, this widens the class of regression models (and statistical models) that can be analyzed with powerful statistical inference tools.

1.3.1 Likelihood, score, and observed information functions

Let y_{obs} be the observed response and $\mathcal{M} = \{p(y; \omega) : y \in \mathcal{Y}, \omega \in \Omega\}$ its statistical model. The definition of the statistical model implies that for every point ω of the parameter space Ω the model function p defines a probability distribution in the sample space $\mathcal{Y}$. This probability distribution describes the assumed random variation of response in the sense that the observed response y_{obs} is only one among the possible observations and various observations occur according to the given distribution. In likelihood-based statistical inference it is important to consider the model function for fixed response as the function of the parameter. In the following we study various concepts connected with this alternative viewpoint. The relevance of these concepts for statistical inference is also discussed.

For the sake of generality let us denote by A the event describing what has been observed with respect to the response. Usually A consists of one point of the sample space, but in certain applications A is a larger subset of the sample space. The function $\Pr(A; \omega)$ of the parameter ω, that is, the probability of the observed event A with respect to the distribution of the statistical model is called the *likelihood function* of the model at the observed event A and is denoted by $L_{\mathcal{M}}(\omega; A)$. If the statistical model consists of discrete distributions and the observed event has the form $A = \{y_{\text{obs}}\}$, the likelihood function is simply the point probability function $p(y_{\text{obs}}; \omega)$ as a function of ω. If, however, the statistical model consists of absolutely continuous distributions, the measurement accuracy of the response has to be taken into account and so the event has the form $A = \{y : y_{\text{obs}} - \delta < y < y_{\text{obs}} + \delta\}$. Assuming for simplicity that the response is one-dimensional, the likelihood function is

$$
\begin{aligned}
L_{\mathcal{M}}(\omega; y_{\text{obs}}) &= \Pr(A; \omega) \\
&= \Pr(y_{\text{obs}} - \delta < y < y_{\text{obs}} + \delta; \omega) \\
&\approx p(y_{\text{obs}}; \omega) 2\delta,
\end{aligned}
$$

which depends on ω only through the density function. Suppose that the response is transformed by a smooth one-to-one function h and denote $z_{\mathrm{obs}} = h(y_{\mathrm{obs}})$. Let g be the inverse of h. Then with respect to the transformed response

$$
\begin{aligned}
L_{\mathcal{M}}(\omega; z_{\mathrm{obs}}) &= \Pr(z_{\mathrm{obs}} - \delta < z < z_{\mathrm{obs}} + \delta) \\
&= \Pr(g(z_{\mathrm{obs}} - \delta) < y < g(z_{\mathrm{obs}} + \delta)) \\
&\approx p(y_{\mathrm{obs}}; \omega) 2\, |g'(h(y_{\mathrm{obs}}))|\, \delta,
\end{aligned}
$$

which differs from the previous expression, but again depends on ω only through the original density function. From this follows that in the absolutely continuous case the likelihood function is up to a constant approximately equal to the density function considered as the function of the parameter. The definition of the likelihood function which applies exactly to the discrete case and approximately to the absolute continuous case is

$$
L_{\mathcal{M}}(\omega; y_{\mathrm{obs}}) = c(y_{\mathrm{obs}}) p(y_{\mathrm{obs}}; \omega), \tag{1.3}
$$

where c is an arbitrary positive function. The above discussion generalizes to models with higher dimensional sample spaces. There are, however, situations in which the above approximation may fail (Lindsey 1996, pp. 75-80). In these situations it is wise to use the exact likelihood function, which is based on the actual measurement accuracy.

Occasionally the event A is more complicated. For example it might be known only that the response is greater than some given value y_{obs}. Then $A = \{y : y > y_{\mathrm{obs}}\}$ and

$$
L_{\mathcal{M}}(\omega; y_{\mathrm{obs}}) = \Pr(y > y_{\mathrm{obs}}; \omega) = 1 - F(y_{\mathrm{obs}}; \omega).
$$

For theoretical reasons, which will be discussed later, instead of the likelihood function it is better to use *logarithmic likelihood function* or *log-likelihood function*; that is, the function

$$
\ell_{\mathcal{M}}(\omega; y_{\mathrm{obs}}) = \ln(L_{\mathcal{M}}(y_{\mathrm{obs}}; \omega)). \tag{1.4}
$$

Often important information about the behavior of likelihood and log-likelihood functions can be obtained from first and second order derivatives of the log-likelihood function with respect to components of the parameter vector. The vector of first derivatives is called *score function* and the negative of the matrix of second derivatives is called *observed information function*. Both practically and theoretically the point in parameter space that maximizes value of likelihood is of special importance. Knowledge of behavior of the likelihood function near this maximum point is vital in likelihood-based inference and for smooth enough likelihood functions the form of the likelihood in the neighborhood of the maximum point can be determined almost completely from knowledge of the maximum point and values of the likelihood,

TABLE 1.3:   Leukemia data.

Treatment	Time of remission
Drug	6+, 6, 6, 6, 7, 9+, 10+, 10, 11+, 13, 16, 17+, 19, 20+, 22, 23, 25+, 32+, 32+, 34+, 35+
Control	1, 1, 2, 2, 3, 4, 4, 5, 5, 8, 8, 8, 8, 11, 11, 12, 12, 15, 17, 22, 23

Time of remission

FIGURE 1.4:   Leukemia data: Times of remission in leukemia data. Symbol ∘ denotes times of the control group, □ denotes uncensored times, and ∎ denotes censored times of the drug group.

score, and observed information functions at this maximum point. In most cases the global maximum point is unique so that score function vanishes and observed information function is positive definite at it. Positive definiteness in itself means that the global maximum point is also the local maximum point.

Example 1.7 Leukemia data

Data in Table 1.3 show times of remission (i.e., freedom from symptoms in a precisely defined sense) of leukemia patients, some patients being treated with the drug 6-mercaptopurine (6-MP), the others serving as a control (Cox and Oakes 1984, p. 7). The columns of data matrix contain values of treatment and time of remission. Censored times have been denoted by a + sign. Figure 1.4 shows values of times of remission.

As an example consider the *Drug* group of the leukemia data. Assume that times of remission are a sample from some Weibull distribution with parameter vector $\omega = (\kappa, \lambda)$, where κ is an unknown shape and λ an unknown scale of the Weibull distribution. The response vector consisting of the first

10 times of remission in the *Drug* group is

$$((6, \infty), 6, 6, 6, 7, (9, \infty), (10, \infty), 10, (11, \infty), 13),$$

where the censored times of remission y_i have been replaced with intervals (y_i, ∞).

Density function of Weibull distribution has the form

$$\kappa \lambda^{-\kappa} y^{\kappa-1} e^{-\left(\frac{y}{\lambda}\right)^{\kappa}}$$

and cumulative distribution function has the form

$$1 - e^{-\left(\frac{y}{\lambda}\right)^{\kappa}}$$

with sample space being $\mathcal{Y} = (0, \infty)$ and parameter space $\Omega = (0, \infty) \times (0, \infty)$, respectively. Thus, the likelihood function for the second observation in the *Drug* group is

$$\kappa \lambda^{-\kappa} 6^{\kappa-1} e^{-\left(\frac{6}{\lambda}\right)^{\kappa}}$$

and the log-likelihood function is

$$\ln(\kappa) - \kappa \ln(\lambda) + (\kappa - 1)\ln(6) - \left(\frac{6}{\lambda}\right)^{\kappa}.$$

For the first observation in the *Drug* group the likelihood function is

$$e^{-\left(\frac{6}{\lambda}\right)^{\kappa}}$$

and the log-likelihood function is

$$-\left(\frac{6}{\lambda}\right)^{\kappa}.$$

The statistical model $\mathcal{M}10$ for the response consisting of the first 10 times of remission is denoted by

$$\mathcal{M}10 = \text{SamplingModel}[\text{WeibullModel}[\kappa, \lambda], 10]$$

with sample space $\mathcal{Y} = (0, \infty)^{10}$, parameter space $\Omega = (0, \infty) \times (0, \infty)$, likelihood function

$$\kappa^6 \lambda^{-6\kappa} 196560^{\kappa-1} e^{-4\left(\frac{6}{\lambda}\right)^{\kappa} - \left(\frac{7}{\lambda}\right)^{\kappa} - \left(\frac{9}{\lambda}\right)^{\kappa} - 2\left(\frac{10}{\lambda}\right)^{\kappa} - \left(\frac{11}{\lambda}\right)^{\kappa} - \left(\frac{13}{\lambda}\right)^{\kappa}},$$

and log-likelihood function

$$6\ln(\kappa) - 6\kappa \ln(\lambda) + (\kappa - 1)\ln(196560)$$
$$- 4\left(\frac{6}{\lambda}\right)^{\kappa} - \left(\frac{7}{\lambda}\right)^{\kappa} - \left(\frac{9}{\lambda}\right)^{\kappa} - 2\left(\frac{10}{\lambda}\right)^{\kappa} - \left(\frac{11}{\lambda}\right)^{\kappa} - \left(\frac{13}{\lambda}\right)^{\kappa}.$$

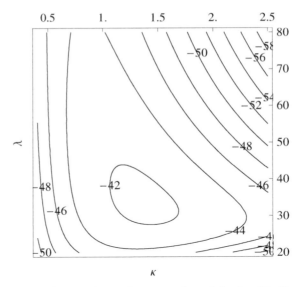

FIGURE 1.5: Leukemia data: Contour plot of the log-likelihood function.

The statistical model $\mathcal{M}$ for times of remission of the whole *Drug* group is denoted by

$$\mathcal{M} = \mathrm{SamplingModel[WeibullModel}[\kappa, \lambda], 21].$$

Figure 1.5 gives a contour plot of the observed log-likelihood function for the whole *Drug* group. This plot seems to indicate that the likelihood function has a maximum value near the point $(1.3, 35)$. The value of the likelihood function at the parameter vector $(1.3, 35)$ for the whole *Drug* group is 7.98527×10^{-19} and that of the log-likelihood function is -41.6715.

The score function for the second observation in the *Drug* group is

$$\begin{pmatrix} -\ln(6)\left(\frac{6}{\lambda}\right)^{\kappa} - \ln\left(\frac{1}{\lambda}\right)\left(\frac{6}{\lambda}\right)^{\kappa} + \ln(6) - \ln(\lambda) + \frac{1}{\kappa} \\ 6^{\kappa}\kappa\left(\frac{1}{\lambda}\right)^{\kappa+1} - \frac{\kappa}{\lambda} \end{pmatrix}$$

and that of the first observation in the *Drug* group is

$$\begin{pmatrix} -6^{\kappa}\ln(6)\left(\frac{1}{\lambda}\right)^{\kappa} - 6^{\kappa}\ln\left(\frac{1}{\lambda}\right)\left(\frac{1}{\lambda}\right)^{\kappa} \\ 6^{\kappa}\kappa\left(\frac{1}{\lambda}\right)^{\kappa+1} \end{pmatrix}.$$

Figure 1.6 shows curves where components of the score function of the whole *Drug* group vanish. Thus, both of the components of the score function seem to vanish near $(1.35, 33.5)$. The value of the score function at the parameter vector $(1.35, 33.5)$ for the *Drug* group is $(0.07284, 0.004466)$.

The value of the observed information function at the parameter vector $(1.35, 33.5)$ is

$$\begin{pmatrix} 8.70192 & 0.150722 \\ 0.150722 & 0.014929 \end{pmatrix}.$$

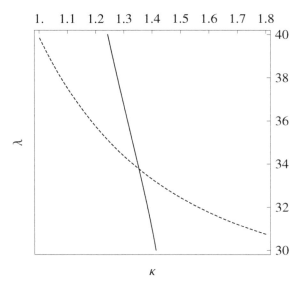

FIGURE 1.6: Leukemia data: Curves on which components of score function vanish. Continuous and dashed lines are for derivatives with respect to κ and λ, respectively.

Eigenvalues of this matrix are equal to 8.70454 and 0.0123147, which means that the observed information matrix is positive definite. This indicates that the likelihood function has at least local maximum near the point $(1.35, 33.5)$.
□

1.3.2 Law of likelihood and relative likelihood function

Let y_{obs} be the observed response and $\mathcal{M} = \{p(y; \omega) : y \in \mathcal{Y},\ \omega \in \Omega\}$ its statistical model.

Law of Likelihood. An observed event concerning the response y_{obs} supports the value ω_1 of the parameter vector more than the value ω_2, if and only if $L_{\mathcal{M}}(\omega_1; A) > L_{\mathcal{M}}(\omega_2; A)$. The ratio $\frac{L_{\mathcal{M}}(\omega_1; A)}{L_{\mathcal{M}}(\omega_2; A)}$ measures how much more statistical evidence supports the value ω_1 of the parameter vector than the value ω_2.

According to the law of likelihood, the likelihood ratio $\frac{L(\omega_1; y_{\mathrm{obs}})}{L(\omega_2; y_{\mathrm{obs}})}$ measures how much more or less the observed response y_{obs} supports the value ω_1 of the parameter vector compared to the value ω_2. Because the likelihood ratio does not change, if the likelihood function is multiplied by a number not depending on the parameter vector ω, it is convenient for the sake of various comparisons to multiply it by an appropriate number. A number generally used is the reciprocal of the maximum value of the likelihood function. The version of the likelihood function obtained in this way is called the *relative (standardized) likelihood function* and it has the form

$$R(\omega; y_{\text{obs}}) = \frac{L(\omega; y_{\text{obs}})}{L(\hat{\omega}; y_{\text{obs}})}, \tag{1.5}$$

where $\hat{\omega}$ is the value of the parameter vector maximizing the likelihood function. The relative likelihood function takes values between 0 and 1 and its maximum value is 1. *Logarithmic relative likelihood function* or *relative (standardized) log-likelihood function* is the logarithm of the relative likelihood function and thus has the form

$$r(\omega; y_{\text{obs}}) = \ell(\omega; y_{\text{obs}}) - \ell(\hat{\omega}; y_{\text{obs}}). \tag{1.6}$$

The relative log-likelihood function has its values in the interval $(-\infty, 0)$ and its maximum value is 0.

Example 1.8 Stressful events data

In a study on the relationship between life stresses and illness one randomly chosen member of each randomly chosen household in a sample from Oakland, California, was interviewed. In a list of 41 events, respondents were asked to note which had occurred within the last 18 months. The results given in Table 1.4 are those recalling only one such stressful event (Lindsey 1996, p. 50; Haberman 1978, p. 3). The columns of data matrix in Table 1.4 contain values of the number of months and count. Figure 1.7 gives plots of counts and logarithms of counts against the month. Consider for the sake of illustration that the observed counts can be modeled as a sample from some Poisson model with unknown mean μ. The statistical model $\mathcal{M}$ for counts is

$$\mathcal{M} = \text{SamplingModel}[\text{PoissonModel}[\mu], 18].$$

Log-likelihood, score, and observed information functions have the following forms:

$$147 \ln(\mu) - 18\mu,$$

$$\frac{147}{\mu} - 18,$$

and

$$\frac{147}{\mu^2},$$

respectively. From these it is easily determined that the value of the parameter maximizing the likelihood function is $147/18 = 49/6$. The relative likelihood function is now

$$5.89792 \times 10^{-71} \mu^{147} e^{-18\mu}$$

TABLE 1.4:
Stressful events
data.

Month	Count
1	15
2	11
3	14
4	17
5	5
6	11
7	10
8	4
9	8
10	10
11	7
12	9
13	11
14	3
15	6
16	1
17	1
18	4

and the relative log-likelihood function is

$$147 \ln(\mu) - 18\mu - 147 \ln\left(\frac{49}{6}\right) + 147.$$

Figure 1.8 gives their graphs.   ⬜

1.4 Likelihood-based methods

> For all purposes, and more particularly for the *communication* of the relevant evidence supplied by a body of data, the values of the Mathematical Likelihood are better fitted to analyze, summarize, and communicate statistical evidence of types too weak to supply true probability statements; it is important that the likelihood always exists, and is directly calculable (Fisher 1956, p. 72).

The law of likelihood provides an ordering among different parameter values, that is, explanations or theories. Thus, it tells us what kind of statements about the unknown parameter value are meaningful. In addition to the above

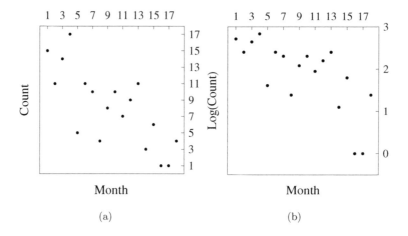

FIGURE 1.7: Stressful events data: Scatter plot of (a) count against month and (b) logarithm of count against month.

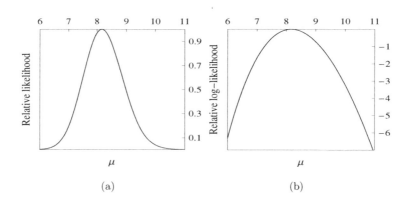

FIGURE 1.8: Stressful events data: Plots of the likelihood of stressful events data. (a) Relative likelihood of μ and (b) relative log-likelihood of μ.

quotation from *Statistical Methods and Scientific Inference* (p. 72) Fisher also notes that

> In the case under discussion a simple graph of the values of the Mathematical Likelihood expressed as a percentage of its maximum, against the possible values of the parameter p, shows clearly enough what values of the parameter have likelihood comparable with the maximum, and outside what limits the likelihood falls to levels at which the corresponding values of the parameter become implausible.

The meaningful statements about the unknown parameter value are thus statements that the unknown parameter value belongs to a subset of parameter space containing all parameter values, the relative likelihood of which is not less than a given number, that is, so-called *likelihood regions*. In this book only likelihood-based region estimates have been used. The uncertainties of these regions have always been calculated from appropriate distributions based on the repeated sampling principle.

1.4.1   Likelihood region

The *pure likelihood approach* gives statistical inference procedures for which the collection of evidential statements is based wholly on the likelihood function. In a sense the likelihood function as a whole is often considered as the evidential statement and giving the evidential meaning of the statistical evidence. More generally, however, the likelihood function is summarized using some simpler evidential statements.

The most common way of summarizing the pure likelihood inference is to use the law of likelihood and, based on that, to give likelihood regions of appropriate or all percentage values. This method was enforced in Fisher (1956) and consists of sets of the form

$$
\begin{aligned}
\mathcal{A}_c &= \{\omega : R(\omega; y_{\text{obs}}) \geq c\} \\
&= \left\{\omega : \frac{L(\omega; y_{\text{obs}})}{L(\hat{\omega}; y_{\text{obs}})} \geq c\right\} \\
&= \{\omega : r(\omega; y_{\text{obs}}) \geq \ln(c)\}
\end{aligned} \tag{1.7}
$$

for some or all $0 < c < 1$. The set $\mathcal{A}_c$ is called $100c\%$ likelihood region and the corresponding evidential statement has the form "The parameter vector ω belongs to the set $\mathcal{A}_c$." According to the law of likelihood, the observed response supports every value of the parameter vector in the likelihood region more than any value outside the region. For each value ω of the parameter vector in the likelihood region one can also say that there does not exist any value of the parameter vector that would be supported by the statistical evidence more than $\frac{1}{c}$-fold compared to ω.

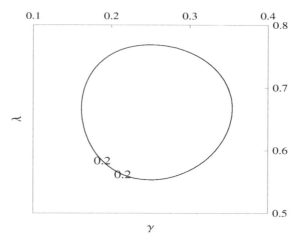

FIGURE 1.9: Respiratory disorders data: Contour plot of 20% likelihood region of respiratory disorders data.

Example 1.9 Respiratory disorders data (cont'd)

As an example consider the respiratory disorders data. Assume that the counts of favorable cases can be modeled as observations from two binomial models, in other words, let us use the following statistical model

$$\mathcal{M} = \text{IndependenceModel}[\{\text{BinomialModel}[64, \gamma], \text{BinomialModel}[60, \lambda]\}].$$

Parameter estimates are equal to $1/4$ and $2/3$, respectively.
The 20% likelihood region has the form

$$\left\{(\gamma, \lambda) : 1.64473 \times 10^{32} \gamma^{16} (1 - \gamma)^{48} \lambda^{40} (1 - \lambda)^{20} \geq 0.2\right\}$$

and Figure 1.9 gives its contour plot. According to the interpretation of the likelihood region, every parameter vector in the likelihood region has the property that there does not exist other parameter vectors with likelihood more than five times as great as the likelihood of the parameter vector in the region.

1.4.2 Uncertainty of likelihood region

In the *likelihood approach* the collection of evidential statements is formed as in the pure likelihood approach. In addition, the uncertainties of the evidential statements are calculated from hypothetical repetitions according to the repeated sampling principle.

For a given parameter vector ω the evidential statement "Parameter vector ω belongs to the $100c\%$ likelihood region" is false if the response y is such that

$$r(\omega; y) < \ln(c).$$

The probability of this event calculated at the given value w of the parameter vector is used as a measure of uncertainty or risk of the likelihood region provided that the probability has the same value or at least approximately the same value for all possible values of the parameter vector.

Example 1.10 Fuel consumption data (cont'd)

For example, assume that in fuel consumption data the amounts of fuel used can be modeled as a random sample from some normal model with known variance $\sigma_0^2 = 10$. The statistical model $\mathcal{M}$ is

$$\mathcal{M} = \text{SamplingModel}[\text{NormalModel}[\mu, \sqrt{10}], 10].$$

The log-likelihood function can be written in the following form:

$$\ell(\mu; y) = \sum_{j=1}^{n} -\frac{(y_j - \mu)^2}{2\sigma_0^2}$$

$$= -\sum_{j=1}^{n} \frac{(y_j - \bar{y} + \bar{y} - \mu)^2}{2\sigma_0^2}$$

$$= -\sum_{j=1}^{n} \frac{(y_j - \bar{y})^2}{2\sigma_0^2} - \frac{n(\bar{y} - \mu)^2}{2\sigma_0^2}.$$

Because the sum is independent of the parameter it can be dropped from the log-likelihood function and so we get

$$\ell(\mu; y) = -\frac{n(\bar{y} - \mu)^2}{2\sigma_0^2}.$$

Clearly this achieves its maximum value at $\mu = \bar{y}$. Thus, the relative log-likelihood function has the same form and $100c\%$ likelihood region is the interval

$$\{\mu \in \mathbb{R} : r(\mu; y) \geq \ln(c)\} = \left\{\mu \in \mathbb{R} : -\frac{n(\bar{y} - \mu)^2}{2\sigma_0^2} \geq \ln(c)\right\}$$

$$= \left(\bar{y} - \sqrt{2\ln\left(\frac{1}{c}\right)}\frac{\sigma_0}{\sqrt{n}}, \bar{y} + \sqrt{2\ln\left(\frac{1}{c}\right)}\frac{\sigma_0}{\sqrt{n}}\right).$$

The number

$$\Pr\left(\{y : r(\mu; y) < \ln(c)\}; \mu\right) = \Phi\left(-\sqrt{2\ln\left(\frac{1}{c}\right)}\right) + 1 - \Phi\left(\sqrt{2\ln\left(\frac{1}{c}\right)}\right)$$

$$= 2\Phi\left(-\sqrt{2\ln\left(\frac{1}{c}\right)}\right)$$

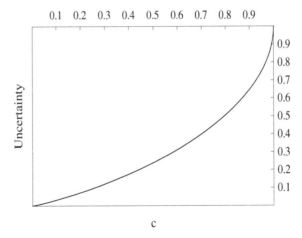

FIGURE 1.10: Fuel consumption data: Plot of the uncertainty as the function of the level c.

can be used as the measure of uncertainty of the likelihood region because it does not depend on μ. Figure 1.10 gives the plot of the uncertainty as the function of the level c and the uncertainties for levels 0.1, 0.2, ..., and 0.9 are 0.032, 0.073, 0.121, 0.176, 0.239, 0.312, 0.398, 0.504, and 0.646, respectively. The level of the likelihood region whose uncertainty is 0.05 is $c = 0.1465$. □

1.5 Profile likelihood-based confidence intervals

> Most practical problems of parametric inference, it seems to me,
> fall in the category of "inference in the presence of nuisance pa-
> rameters," with the acknowledgement that various functions $g(\theta)$,
> which might be components of a parameter vector, are simultane-
> ous interest. That is, in practice, we may wish to make inferences
> about each of many such functions $g(\theta)$ in turn, keeping in mind
> that joint inferences are being made, yet requiring separate state-
> ments for ease of contemplation (Kass in Reid 1988, p. 236).

As R.E. Kass says in the above quotation one usually is interested in one or more real-valued functions of the parameter vector and wants to make inferences on the unknown values of those functions. In this book the main emphasis is on construction of a confidence interval or a set of confidence intervals for the unknown value of a given interest function.

The above quotation does not exactly say how the likelihood should be used. A natural way that is in agreement with Fisher's ideas on likelihood and with the above statement is provided by the so-called profile likelihood function.

1.5.1   Profile likelihood function

Let y_{obs} be the observed response and $\mathcal{M} = \{p(y; \omega) : y \in \mathcal{Y}, \omega \in \Omega\}$ its statistical model. In most practical problems only part of the parameter vector or, more generally, the value of a given function of the parameter vector is of interest.

Let $g(\omega)$ be a given interest function with q-dimensional real vector ψ as the value. Then the function

$$L_g(\psi; y_{\text{obs}}) = \max_{\{\omega \in \Omega : g(\omega) = \psi\}} L(\omega; y_{\text{obs}}) = L(\tilde{\omega}_\psi; y_{\text{obs}}) \qquad (1.8)$$

is called the *profile likelihood function* of the interest function g induced by the statistical evidence $(y_{\text{obs}}, \mathcal{M})$. The value $\tilde{\omega}_\psi$ of the parameter vector maximizes the likelihood function in the subset $\{\omega \in \Omega : g(\omega) = \psi\}$ of the parameter space. The function

$$\begin{aligned} \ell_g(\psi; y_{\text{obs}}) &= \max_{\{\omega \in \Omega : g(\omega) = \psi\}} \ell(\omega; y_{\text{obs}}) \\ &= \ell(\tilde{\omega}_\psi; y_{\text{obs}}) \\ &= \ln(L_g(\psi; y_{\text{obs}})) \end{aligned} \qquad (1.9)$$

is called the *logarithmic profile likelihood function* or *profile log-likelihood function* of the interest function g induced by the statistical evidence $(y_{\text{obs}}, \mathcal{M})$. Furthermore, functions

$$R_g(\psi; y_{\text{obs}}) = \frac{L_g(\psi; y_{\text{obs}})}{L_g(\hat{\psi}; y_{\text{obs}})} = \frac{L_g(\psi; y_{\text{obs}})}{L(\hat{\omega}; y_{\text{obs}})} \qquad (1.10)$$

and

$$\begin{aligned} r_g(\psi; y_{\text{obs}}) &= \ell_g(\psi; y_{\text{obs}}) - \ell_g(\hat{\psi}; y_{\text{obs}}) \\ &= \ell_g(\psi; y_{\text{obs}}) - \ell(\hat{\omega}; y_{\text{obs}}) \end{aligned} \qquad (1.11)$$

are called the *relative profile likelihood function* and the *logarithmic relative profile likelihood function* or *relative profile log-likelihood function* of interest function g, respectively. Because in this book no other likelihood functions than actual likelihood functions of the whole parameter and profile likelihood functions of various interest functions g are considered, the phrases (*relative*) *likelihood function* and (*relative*) *log-likelihood function* in the former case and (*relative*) *likelihood function* and (*relative*) *log-likelihood function* of g in the latter case are used.

When the interest function g is real valued, the parameter vectors $\tilde{\omega}_\psi$ form a curve in the parameter space. This curve is called the *profile curve* of the

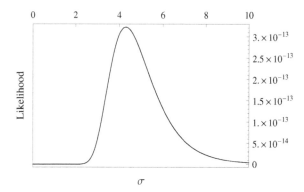

FIGURE 1.11: Fuel consumption data: Plot of the likelihood function of σ for a given value 15 of μ.

interest function g. If $h(\omega)$ is some other real valued function of the parameter vector, the function $h(\tilde{\omega}_\psi)$ of ψ is called the *profile trace* of h with respect to g.

Example 1.11 Fuel consumption data (cont'd)

Assume that the fuel consumption measurements may be modeled as a sample from some normal distribution, that is, with statistical model

$$\mathcal{M} = \text{SamplingModel}[\text{NormalModel}[\mu, \sigma], 10].$$

Assume also that the mean of the normal distribution is the parameter of interest.

The likelihood function of the statistical model is

$$\frac{e^{\frac{-5.\mu^2 + 177.35\mu - 1627.48}{\sigma^2}}}{32\pi^5\sigma^{10}}.$$

To find the likelihood function of the mean μ we need to maximize the likelihood function with respect to the standard deviation σ; for example, in case the mean has the value of 15. Figure 1.11 gives the plot of the likelihood function as the function of σ.

Now, in general, if the response y_{obs} consists of n real numbers and is assumed to be a sample from some normal distribution with unknown mean and variance the likelihood function as the function of the standard deviation σ has the form

$$(2\pi)^{-\frac{n}{2}}\sigma^{-n}e^{-\frac{SS}{2\sigma^2}},$$

where

$$SS = \sum_{j=1}^{n}(y_j - \mu)^2$$

is the sum of squares with respect to the given value μ of the mean. Assuming that not all the observations have the same value, the sum of squares is positive for all values of μ. Because the value of the standard deviation σ maximizing the likelihood function also maximizes the log-likelihood function, we can consider the function

$$-\frac{n}{2}\ln(2\pi) - n\ln(\sigma) - \frac{SS}{2\sigma^2},$$

the derivative of which is

$$-\frac{n}{\sigma} + \frac{SS}{\sigma^3}.$$

Clearly this derivative takes the value zero at $\sqrt{\frac{SS}{n}}$ and is positive for smaller values of σ and negative for larger values of σ. This means that $\tilde{\sigma}_\mu = \sqrt{\frac{SS}{n}}$ and the likelihood function of μ is

$$L_\mu(\psi; y_{\text{obs}}) = \max_{0<\sigma<\infty} (2\pi)^{-\frac{n}{2}} \sigma^{-n} e^{-\frac{1}{2\sigma^2}\sum_{j=1}^{n}(y_j-\psi)^2}$$

$$= \left(2\pi \frac{\sum_{j=1}^{n}(y_j-\psi)^2}{n}\right)^{-\frac{n}{2}} e^{-\frac{n}{2}}.$$

The sum of squares

$$\sum_{j=1}^{n}(y_j-\psi)^2 = \sum_{j=1}^{n}(y_j-\bar{y})^2 + n(\bar{y}-\psi)^2$$

is minimized when ψ takes the value $\bar{y}$ and so the relative likelihood function of μ has the following form:

$$R_\mu(\psi; y_{\text{obs}}) = \frac{L_\mu(\psi; y_{\text{obs}})}{L_\mu(\bar{y}; y_{\text{obs}})}$$

$$= \left(\frac{\sum_{j=1}^{n}(y_j-\psi)^2}{\sum_{j=1}^{n}(y_j-\bar{y})^2}\right)^{-\frac{n}{2}}$$

$$= \left(1 + \frac{n(\bar{y}-\psi)^2}{\sum_{j=1}^{n}(y_j-\bar{y})^2}\right)^{-\frac{n}{2}}$$

$$= \left(1 + \frac{1}{n-1}T_\psi(y_{\text{obs}})^2\right)^{-\frac{n}{2}},$$

where T_ψ is the t-statistic

$$T_\psi(y) = \frac{\sqrt{n}(\bar{y}-\psi)}{\sqrt{\frac{1}{n-1}\sum_{j=1}^{n}(y_j-\bar{y})^2}} = \frac{\bar{y}-\psi}{\frac{s}{\sqrt{n}}}.$$

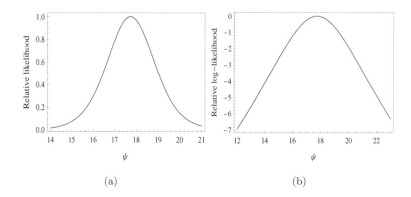

FIGURE 1.12:   Fuel consumption data: Plots of (a) relative likelihood of μ and (b) relative log-likelihood of μ.

The log-likelihood function and the relative log-likelihood function of μ are

$$\ell_\mu\left(\psi; y_{\text{obs}}\right) = -\frac{n}{2}\ln\left(\frac{\sum_{j=1}^{n}\left(y_j - \psi\right)^2}{n}\right) - \frac{n}{2}$$

$$r_\mu\left(\psi; y_{\text{obs}}\right) = -\frac{n}{2}\ln\left(1 + \frac{n\left(\bar{y} - \psi\right)^2}{\sum_{j=1}^{n}\left(y_j - \bar{y}\right)^2}\right)$$

$$= -\frac{n}{2}\ln\left(1 + \frac{1}{n-1}T_\psi(y)^2\right).$$

Figure 1.12 shows plots of relative likelihood and log-likelihood functions of μ. Figure 1.13 illustrates the intuitive idea behind the likelihood function of μ as the silhouette of a structure when viewed from far away. ▯

1.5.2   Profile likelihood region and its uncertainty

With the help of the relative likelihood function of interest function g one can construct the so-called *profile likelihood regions*. The set

$$\{\psi : R_g(\psi; y_{\text{obs}}) \geq c\} = \{\psi : r_g(\psi; y_{\text{obs}}) \geq \ln(c)\} \tag{1.12}$$

of values of the interest function $g(\omega)$ is the $100c\%$ profile likelihood region. The value $\psi = g(\omega)$ of the parameter function does not belong to the $100c\%$ profile likelihood region if the response is such that

$$r_g(\psi; y) < \ln(c).$$

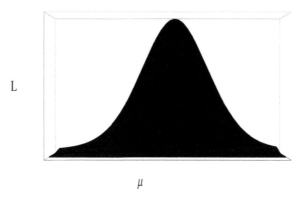

L

μ

FIGURE 1.13: Fuel consumption data: Plot of likelihood function viewed far away from the origin on the negative side of the σ-axis.

The probability of this event, calculated at a given value ω of the parameter vector, is used as a measure of uncertainty of the statement that the unknown value of the interest function belongs to the $100c\%$ profile likelihood region, provided that this probability has the same value or at least approximately the same value for all values of the parameter vector. One minus this probability is called the *confidence level* of the profile likelihood region and the region is called the (approximate) $(1-\alpha)$-level *profile likelihood-based confidence region* for the interest function. Because as is discussed in the next section under mild assumptions concerning the interest function $g(\omega)$ and its statistical model the random variable $-2r_g(\psi; y)$ has approximately the χ^2-distribution with q degrees of freedom, the set

$$\left\{\psi : r_g\left(\psi; y_{\text{obs}}\right) \geq -\frac{\chi^2_{1-\alpha}[q]}{2}\right\} = \left\{\psi : -2r_g\left(\psi; y_{\text{obs}}\right) \leqslant \chi^2_{1-\alpha}[q]\right\}$$

$$= \left\{\psi : \ell_g(\psi) \geq \ell_g(\hat{\psi}) - \frac{\chi^2_{1-\alpha}[q]}{2}\right\} (1.13)$$

is the approximate $(1-\alpha)$-level confidence region for the interest function $g(\omega)$. In some cases the distribution of $-2r_g(\psi; y)$ is exactly the χ^2-distribution with q degrees of freedom and then the set is an exact confidence region. Sometimes the distribution of the random variable $-2r_g(\psi; y)$ is determined by some other known distribution, usually the F-distribution.

Example 1.12 **Fuel consumption data (cont'd)**

Assume that the fuel consumption in the fuel consumption data can be considered as a sample from some normal distribution with unknown mean and variance. Let the mean μ be the parameter function of interest. The statistical model is defined by

$$\mathcal{M} = \text{SamplingModel}[\text{NormalModel}[\mu, \sigma], 10].$$

The $100c\%$ profile likelihood region for the mean μ is

$$\{\psi : r_g(\psi; y_{\text{obs}}) \geq \ln(c)\} = \left\{\psi : -\frac{n}{2} \ln\left(1 + \frac{1}{n-1} T_\psi (y_{\text{obs}})^2\right) \geq \ln(c)\right\}$$
$$= \left\{\psi : T_\psi (y_{\text{obs}})^2 \leqslant (n-1)\left(c^{-\frac{2}{n}} - 1\right)\right\}.$$

Because the t-statistic T_ψ has the t-distribution with $n-1$ degrees of freedom, when the mean μ has the value ψ, the uncertainty of the $100c\%$ profile likelihood region is α, if c is chosen such that

$$(n-1)\left(c^{-\frac{2}{n}} - 1\right) = \left(t^*_{\frac{\alpha}{2}}[n-1]\right)^2.$$

Thus, the interval

$$\left\{\bar{y}_{\text{obs}} - t^*_{\frac{\alpha}{2}}[n-1]\frac{s}{\sqrt{n}}, \bar{y}_{\text{obs}} + t^*_{\frac{\alpha}{2}}[n-1]\frac{s}{\sqrt{n}}\right\}$$

is the $(1-\alpha)$-level confidence interval for the mean μ and its value is $(15.238, 20.232)$. ⌷

1.5.3   Profile likelihood-based confidence interval

For real valued interest functions the profile likelihood-based confidence regions are usually intervals and so those regions are called (approximate) $(1-\alpha)$-level *profile likelihood-based confidence intervals*. Thus, the set

$$\left\{\psi : \ell_g(\psi; y_{\text{obs}}) \geq \ell_g(\hat{\psi}; y_{\text{obs}}) - \frac{\chi^2_{1-\alpha}[1]}{2}\right\} = [\hat{\psi}_L, \hat{\psi}_U] \qquad (1.14)$$

forms the (approximate) $(1-\alpha)$-level profile likelihood-based confidence interval for the interest function $\psi = g(\omega)$. The statistics $\hat{\psi}_L$ and $\hat{\psi}_U$ are called the *lower* and *upper limits* of the confidence interval, respectively.

Example 1.13 Respiratory disorders data (cont'd)

Let us consider placebo and test groups of the respiratory disorder data. Assume that counts of favorable cases in both groups may be modeled as observations from binomial distributions and define the statistical model

$$\mathcal{M} = \text{IndependenceModel}[\{\text{BinomialModel}[64, \gamma], \text{BinomialModel}[60, \lambda]\}],$$

where γ is the probability of a favorable case in the placebo group and λ is that in the test group.

TABLE 1.5:   Respiratory disorders data: Estimates and approximate 0.95-level confidence intervals of interest functions.

Parameter function	Estimate	Lower limit	Upper limit
γ	0.250	0.155	0.365
λ	0.667	0.542	0.777
ψ_1	6.000	2.803	13.411
ψ_2	41670	25000	56690

Assume that parameter functions of interest are the odds ratio

$$\psi_1 = \frac{\lambda(1-\gamma)}{(1-\lambda)\gamma}$$

and a potential increase of favorable cases among $100,000$ people having respiratory disorders, that is,

$$\psi_2 = 100000(\lambda - \gamma).$$

Table 1.5 gives estimates and approximate 0.95-level profile likelihood-based confidence intervals for model parameters and interest functions. Figure 1.14 contains plots of corresponding log-likelihood functions. ◻

Example 1.14 Leukemia data (cont'd)

Assume that the times of remission can be considered as samples from two exponential distributions. The statistical model under these assumptions is

$$\mathcal{M} = \text{IndependenceModel}\left[\{\text{SamplingModel}\left[\text{ExponentialModel}\left[\mu_1\right], 21\right],\right.$$
$$\left.\text{SamplingModel}\left[\text{ExponentialModel}\left[\mu_2\right], 21\right]\}\right].$$

Of interest in addition to model parameters might be the difference of means or their ratio, that is, parameter functions

$$\psi_1 = \mu_1 - \mu_2$$

and

$$\psi_2 = \mu_1/\mu_2,$$

but a better characteristic for describing the difference of distributions might be the probability that a random observation from the first distribution would be greater than a random statistically independent observation from the second distribution. This interest function is related to the so-called *Receiver Operating Characteristic* (ROC) curve, namely, it is area under this curve (AUC). Under current assumptions this interest function has the form

$$\psi_3 = \frac{\mu_1}{\mu_1 + \mu_2}.$$

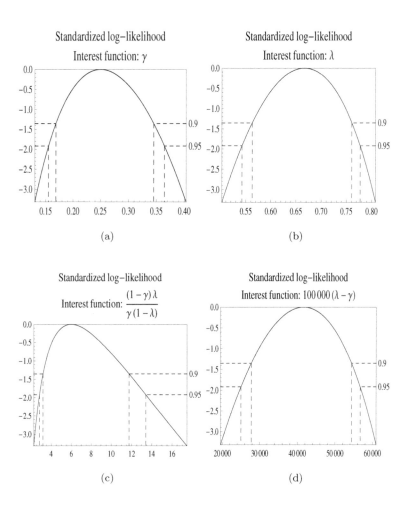

FIGURE 1.14:   Respiratory disorders data: Plots of log-likelihoods of (a) γ, (b) λ, (c) ψ_1, and (d) ψ_2.

TABLE 1.6:   Leukemia data: Estimates and approximate 0.95-level confidence intervals of interest functions.

Parameter function	Estimate	Lower limit	Upper limit
μ_1	39.889	22.127	83.013
μ_2	8.667	5.814	13.735
ψ_1	31.222	12.784	74.444
ψ_2	4.603	2.174	10.583
ψ_3	0.822	0.685	0.914
ψ_4	-2.820	-8.677	-0.634

Another parameter function describing the difference of the distributions is the so-called *Kullback-Leibler divergence*, which under current assumptions is

$$\psi_4 = 2 - \frac{\mu_2}{\mu_1} - \frac{\mu_1}{\mu_2}.$$

Table 1.6 gives maximum likelihood estimates and approximate 0.95-level profile likelihood-based confidence intervals for parameters and interest functions.

Figure 1.15 shows graphs of log-likelihoods of parameters and interest functions. □

1.5.4   Calculation of profile likelihood-based confidence intervals

If the (approximate) $(1 - \alpha)$-level profile likelihood-based confidence region of the real valued interest function $\psi = g(\omega)$ is an interval, its end points $\hat{\psi}_L$ and $\hat{\psi}_U$ satisfy the relations $\hat{\psi}_L < \psi < \hat{\psi}_U$ and

$$\ell_g(\hat{\psi}_L) = \ell_g(\hat{\psi}_U) = \ell_g(\hat{\psi}) - \frac{\chi^2_{1-\alpha}[1]}{2}. \tag{1.15}$$

Thus, $\hat{\psi}_L$ and $\hat{\psi}_U$ are roots of the equation $\ell_g(\psi) = \ell_g(\hat{\psi}) - \frac{\chi^2_{1-\alpha}[1]}{2}$. Consequently most applications of the profile likelihood-based intervals have determined $\hat{\psi}_L$ and $\hat{\psi}_U$ using some iterative root finding method. This approach involves the solution of an optimization problem in every trial value and, depending on the method, also the calculation of the derivatives of the logarithmic profile likelihood function at the same trial value.

The approximate $(1-\alpha)$-level profile likelihood confidence set for parameter function $\psi = g(\omega)$ satisfies the relation

$$\left\{ \psi : \ell_g(\psi) > \ell_g(\hat{\psi}) - \frac{\chi^2_{1-\alpha}[1]}{2} \right\} = \{g(\omega) : \omega \in \mathcal{R}_{1-\alpha}(y_{\text{obs}})\}, \tag{1.16}$$

where

$$\mathcal{R}_{1-\alpha}(y_{\text{obs}}) = \left\{ \omega : \ell(\omega) > \ell(\hat{\omega}) - \frac{\chi^2_{1-\alpha}[1]}{2} \right\} \tag{1.17}$$

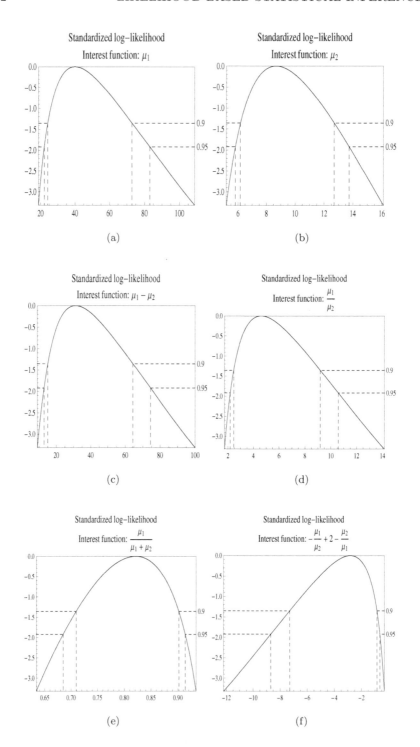

FIGURE 1.15: Leukemia data: Plots of log-likelihoods of interest functions (a) μ_1, (b) μ_2, (c) ψ_1, (d) ψ_2, (e) ψ_3, and (f) ψ_4.

is a likelihood region for the whole parameter vector. This result is true because the real number ψ belongs to the profile likelihood region if and only if there exists a parameter vector ω^* such that $g(\omega^*) = \psi$ and

$$\sup_{\{\omega \in \Omega : g(\omega) = \psi\}} \ell(\omega) = \ell_g(\psi) > \ell(\omega^*) > \ell_g(\hat{\psi}) - \frac{\chi^2_{1-\alpha}[1]}{2}.$$

That is, if and only if there exists a parameter vector ω^* belonging to the likelihood region $\mathcal{R}_{1-\alpha}(y_{\text{obs}})$ such that it satisfies the equation $g(\omega^*) = \psi$. So the number ψ belongs to the profile likelihood region if and only if it belongs to the set $\{g(\omega) : \omega \in \mathcal{R}_{1-\alpha}(y_{\text{obs}})\}$.

Assume now that the profile likelihood-based confidence region is an interval. Then the end points $\hat{\psi}_L$ and $\hat{\psi}_U$ satisfy the following relations:

$$\hat{\psi}_L = \inf_{\omega \in \mathcal{R}_{1-\alpha}(y_{\text{obs}})} g(\omega) \text{ and } \hat{\psi}_U = \sup_{\omega \in \mathcal{R}_{1-\alpha}(y_{\text{obs}})} g(\omega). \tag{1.18}$$

Now let the likelihood region be a bounded and connected subset of the parameter space. If the log-likelihood and the interest functions are continuous in the likelihood region, the latter with non-vanishing gradient, then the profile likelihood-based confidence region is an interval $(\hat{\psi}_L, \hat{\psi}_U)$ with

$$\hat{\psi}_L = g(\tilde{\omega}_L) = \inf_{\ell(\omega) = \ell(\hat{\omega}) - \chi^2_\alpha[1]/2} g(\omega) \tag{1.19}$$

and

$$\hat{\psi}_U = g(\tilde{\omega}_U) = \sup_{\ell(\omega) = \ell(\hat{\omega}) - \chi^2_\alpha[1]/2} g(\omega). \tag{1.20}$$

This follows from the assumptions because they imply that the set $\mathcal{R}_{1-\alpha}(y_{\text{obs}})$ is an open, connected, and bounded subset of the d-dimensional space. Thus, the closure of $\mathcal{R}_{1-\alpha}(y_{\text{obs}})$ is a closed, connected, and bounded set. From the assumptions concerning g it follows that it attains its infimum and supremum on the boundary of $\mathcal{R}_{1-\alpha}(y_{\text{obs}})$ and takes every value between infimum and supremum somewhere in $\mathcal{R}_{1-\alpha}(y_{\text{obs}})$.

Under the above assumptions the solutions of the following constrained minimization (maximization) problem

$$\min(\max) g(\omega) \text{ under } \ell(\omega; y_{\text{obs}}) = \ell(\hat{\omega}; y_{\text{obs}}) - \frac{\chi^2_\alpha[1]}{2} \tag{1.21}$$

gives the lower (upper) limit point of the profile likelihood-based confidence interval. This problem is rarely explicitly solvable and requires use of some kind of iteration (Uusipaikka 1996).

Example 1.15 Respiratory disorders data (cont'd)

Let us consider the placebo and test groups of the respiratory disorders data. Assume that the counts of the favorable cases in both groups may be

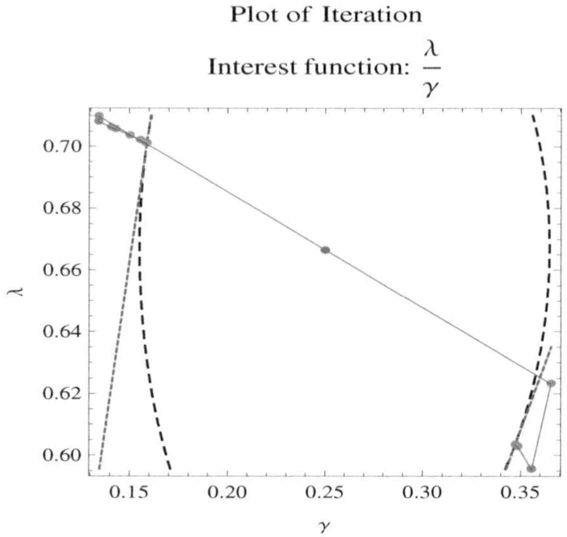

Plot of Iteration

Interest function: $\dfrac{\lambda}{\gamma}$

FIGURE 1.16: Respiratory disorders data: Iteration steps of calculation of the confidence interval for odds ratio.

modeled as observations from binomial distributions and that the parameter function of interest is the ratio $\frac{\lambda}{\gamma}$, where λ and γ are the "success" probabilities in test and placebo groups, respectively; that is,

$$\mathcal{M} = \text{IndependenceModel}[\{\text{BinomialModel}[64, \gamma], \text{BinomialModel}[60, \lambda]\}].$$

Figure 1.16 shows iteration steps in the course of calculation of the confidence interval. ▯

Example 1.16 Leukemia data (cont'd)

Let us consider AUC under ROC. Figure 1.17 shows iteration steps for this interest function. ▯

1.5.5 Comparison with the delta method

Let $g(\omega)$ be a given real valued interest function with ψ as the value. The most often used procedure to construct a confidence interval for ψ is the so-called *delta method*. The delta method interval has the form

$$\psi \in \hat{\psi} \mp z^*_{\alpha/2} \sqrt{\frac{\partial g(\hat{\omega})}{\partial \omega}^{\mathrm{T}} J(\hat{\omega})^{-1} \frac{\partial g(\hat{\omega})}{\partial \omega}}, \tag{1.22}$$

where $\hat{\psi} = g(\hat{\omega})$ is the maximum likelihood estimate of ψ, $J(\omega)$ is the observed information matrix, and $z^*_{\alpha/2}$ is the $(1 - \alpha/2)$-quantile of standard normal

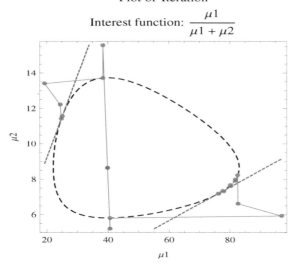

FIGURE 1.17: Leukemia data: Iteration steps of calculation of the confidence interval for AUC under ROC.

distribution. When the interest function consists of a component ω_i of the parameter vector ω, the previous interval takes the form

$$\omega_i \in \hat{\omega}_i \mp z_{\alpha/2}^* s_{\hat{\omega}_i}, \tag{1.23}$$

where the *standard error of estimate*

$$s_{\hat{\omega}_i} = \sqrt{(J(\hat{\omega})^{-1})_{ii}} \tag{1.24}$$

is the square root of the ith diagonal element of the inverse $J(\hat{\omega})^{-1}$. This latter confidence interval is better known by the name *Wald confidence interval*.

Wald confidence interval is derived from the asymptotic normal distribution of the maximum likelihood estimate $\hat{\omega}$, that is,

$$\hat{\omega} \simeq \text{MultivariateNormalModel}[d, \omega, I(\omega)^{-1}], \tag{1.25}$$

where $I(\omega)$ is the Fisher or expected information matrix which can be estimated by the observed information matrix $J(\hat{\omega})$ (see Section 1.7). Thus, first we have from the asymptotic normal distribution

$$\frac{\hat{\omega}_i - \omega_i}{\sqrt{(I(\omega)^{-1})_{ii}}} \simeq \text{NormalModel}[0, 1]$$

and then by inserting the estimate

$$\frac{\hat{\omega}_i - \omega_i}{\sqrt{(J(\hat{\omega})^{-1})_{ii}}} \simeq \text{NormalModel}[0, 1].$$

from which the Wald confidence interval follows.

The delta method interval is then obtained from the Taylor-series expansion

$$g(\hat{\omega}) \approx g(\omega) + \frac{\partial g(\omega)}{\partial \omega}^{\mathrm{T}} (\hat{\omega} - \omega),$$

which gives

$$\hat{\psi} = g(\hat{\omega}) \simeq \text{NormalModel}\left[\psi, \frac{\partial g(\omega)}{\partial \omega}^{\mathrm{T}} I(\omega)^{-1} \frac{\partial g(\omega)}{\partial \omega}\right].$$

So

$$\frac{\hat{\psi} - \psi}{\sqrt{\frac{\partial g(\omega)}{\partial \omega}^{\mathrm{T}} I(\omega)^{-1} \frac{\partial g(\omega)}{\partial \omega}}} \simeq \text{NormalModel}[0, 1],$$

from which by replacing the denominator by its estimator we get

$$\frac{\hat{\psi} - \psi}{\sqrt{\frac{\partial g(\hat{\omega})}{\partial \omega}^{\mathrm{T}} J(\hat{\omega})^{-1} \frac{\partial g(\hat{\omega})}{\partial \omega}}} \simeq \text{NormalModel}[0, 1].$$

It can be shown that Wald and delta method intervals are approximations of profile likelihood-based confidence intervals. These approximations have, however, some serious drawbacks, the most serious of which is their not being invariant with respect to monotone transformations of the interest function. Second, these intervals often include impossible values of the interest function.

Example 1.17 Fuel consumption data (cont'd)

Assume that the fuel consumption measurements may be modeled as a sample from some normal distribution and the 0.95-quantile of the normal distribution is the parameter of interest, that is,

$$\psi = \mu + z_{0.95}\sigma,$$

where $z_{0.95}$ is the 0.95-quantile of the standard normal distribution. Now the maximum likelihood estimate of ψ is 23.182 and approximate 95%-level profile likelihood-based and delta method confidence intervals for ψ are $(20.72, 27.73)$ and $(20.03, 26.33)$, respectively. In a simulation of $10,000$ samples from normal distribution with mean 16.716 and standard deviation 2.822 the actual coverage probabilities were 0.933 and 0.818 for profile likelihood-based and delta method intervals, respectively.

An explanation for the weak performance of the delta method in this case can be seen from Figure 1.18, which shows the log-likelihood function of ψ for the actual data. The graph of log-likelihood function is asymmetric and so the profile likelihood-based confidence interval is also asymmetric with respect to the maximum likelihood estimate $\hat{\psi}$. The delta method interval is

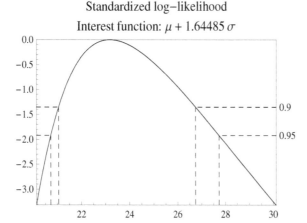

FIGURE 1.18: Fuel consumption data: Plot of log-likelihood of $\psi = \mu + z_{0.95}\sigma$.

forced to be symmetric and because of this it often misses the "true" value of ψ because of a too low upper limit. When applying the delta method it is important to use an appropriate scale for the interest function by transforming first ψ to $\lambda = h(\psi)$ so that the distribution of $\hat{\lambda} = h(\hat{\psi})$ is better approximated by normal distribution, especially with approximately constant variance. Then using the delta method interval (λ^L, λ^U) for λ a better interval $(h^{-1}(\lambda^L), h^{-1}(\lambda^U))$ for ψ is obtained. When applying the profile likelihood method there is no need, however, to make any transformations because of the invariance of profile likelihood-based intervals. In a sense the profile likelihood method automatically chooses the best transformation and the user need not worry about that (Pawitan 2001, p. 47). □

1.6 Likelihood ratio tests

> The level of significance in such cases fulfils the conditions of a measure of the rational grounds for the disbelief it engenders. It is more primitive, or elemental than, and does not justify, any exact probability statement about the proposition (Fisher 1956, p. 43).

Sometimes the researcher has a study hypothesis that he wants to confirm with observations. Significance tests have been designed for this purpose. In this book the main method of statistical inference is construction of profile likelihood-based confidence intervals for real-valued parameter functions of

TABLE 1.7:
Bacterial concentrations
data.

Time	Concentration
0	31
1	26
2	19
6	15
12	20

interest. Occasionally significance tests will, however, be provided.

1.6.1   Model restricted by hypothesis

Let y_{obs} be the observed response and $\mathcal{M} = \{p(y; \omega) : y \in \mathcal{Y}, \omega \in \Omega\}$ its statistical model. A *statistical hypothesis* is any assumption about the parameter ω of the model. Every hypothesis H defines a subset Ω_{H} of the parameter space Ω of the statistical model $\mathcal{M}$ such that the hypothesis consists of the assumption $\omega \in \Omega_{\text{H}}$. The given hypothesis is often denoted by H $: \omega \in \Omega_{\text{H}}$. Usually the hypothesis is defined either so that for a given vector valued function $g(\omega)$ of the parameter and a given vector ψ one defines $\Omega_{\text{H}} = \{\omega \in \Omega : g(\omega) = \psi\}$, or so that for a given set Θ and a given mapping t from the set Θ to the parameter space Ω one defines $\Omega_{\text{H}} = \{t(\theta) : \theta \in \Theta\}$. The former case is denoted by H $: g(\omega) = \psi$. The statistical model $\mathcal{M}_{\text{H}} = \{p(y; \omega) : y \in \mathcal{Y}, \omega \in \Omega_{\text{H}}\}$ is called the *restricted statistical model* defined by the hypothesis H $: \omega \in \Omega_{\text{H}}$.

The hypothesis H is a *simple hypothesis* if it determines the distribution of the response completely; in other words, if the subset Ω_{H} of the parameter space has only one member ω_{H} or $\Omega_{\text{H}} = \{\omega_{\text{H}}\}$. A simple hypothesis is denoted by H $: \omega = \omega_{\text{H}}$. A hypothesis that is not simple is called a *composite hypothesis*.

Example 1.18   Bacterial concentrations data

These data arose from an investigation of various media to be used for storing micro-organisms in a deep freeze. The results are bacterial concentrations (counts in a fixed area) measured at the initial freezing (-70°C) and then at 1, 2, 6, and 12 months afterward (Francis *et al.* 1993, p. 531; Ludlam *et al.* 1989). The columns of data matrix in Table 1.7 contain values of time and concentration. Figure 1.19 plots natural logarithms of counts against time.

Consider as an example the bacterial concentrations data and assume that bacterial concentrations can be modeled as statistically independent Poisson-distributed counts with possibly different means μ_1, μ_2, μ_3, μ_4, and μ_5.

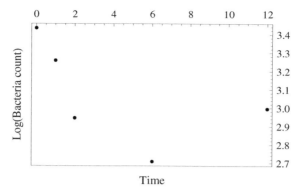

FIGURE 1.19: Bacterial concentrations data: Graph of natural logarithms of counts against time.

The statistical model $\mathcal{M}$ of the counts is

$$\mathcal{M} = \text{IndependenceModel}[\text{PoissonModel}[\mu], 5]$$

or

$$\mathcal{M} = \{p(y; \omega) : y \in \mathcal{Y} = \{0, 1, 2, \ldots\}^5, \omega \in \Omega = (0, \infty)^5\},$$

where $y = (y_1, y_2, y_3, y_4, y_5)$ is the response vector and $\omega = (\mu_1, \ldots, \mu_5)$ is the parameter consisting of the means of the Poisson-distributions and the point probability function has the form

$$p(y; \omega) = \frac{\mu_1^{y_1} \mu_2^{y_2} \mu_3^{y_3} \mu_4^{y_4} \mu_5^{y_5} e^{-\mu_1 - \mu_2 - \mu_3 - \mu_4 - \mu_5}}{y_1! y_2! y_3! y_4! y_5!}.$$

The restricted statistical model defined by the hypothesis H : $\mu_1 = \mu_2 = \mu_3 = \mu_4 = \mu_5$ can be defined in two ways. If t is a mapping from the set $(0, \infty)$ to the parameter space such that $t(\theta) = (\theta, \theta, \theta, \theta, \theta)$ the subset $\Omega_H = \{t(\theta) : \theta \in (0, \infty)\}$ of the parameter space defines the hypothesis H, and the restricted statistical model is

$$\mathcal{S} = \text{Submodel}[\mathcal{M}, \theta, \{\theta, \theta, \theta, \theta, \theta\}]$$

and its point probability function

$$p(y_1, \ldots, y_5; \theta) = \frac{\theta^{y_1 + y_2 + y_3 + y_4 + y_5} e^{-5\theta}}{y_1! y_2! y_3! y_4! y_5!}$$

for all $y_1, \ldots, y_5 \in \{0, 1, 2, \ldots\}$ and $\theta \in (0, \infty)$.   ▯

1.6.2   Likelihood of the restricted model

Let us consider first the situation where the parameter space contains only two elements, that is, $\Omega = \{\omega_1, \omega_2\}$, and we are interested in the simple hypothesis $H : \omega = \omega_1$. According to the law of likelihood, the likelihood ratio $\frac{L(\omega_2; y_{\text{obs}})}{L(\omega_1; y_{\text{obs}})}$ indicates how much more or less the statistical evidence consisting of the data y_{obs} and the model $\mathcal{M}$ supports the parameter value ω_2 compared with the parameter value ω_1 of the hypothesis. A large value of the likelihood ratio is interpreted to support the statement "Hypothesis $H : \omega = \omega_1$ is untrue."

Assume next that the parameter space contains more than two elements and we are still interested in a simple hypothesis $H : \omega = \omega_H$. The likelihood ratio $\frac{L(\omega; y_{\text{obs}})}{L(\omega_H; y_{\text{obs}})}$ measures again how much more or less the statistical evidence supports some other parameter value ω compared with the parameter value ω_H of the hypothesis. In this case, the large value of the ratio

$$\frac{\max\limits_{\omega \in \Omega} L(\omega; y_{\text{obs}})}{L(\omega_H; y_{\text{obs}})}$$

is taken to support the statement "Hypothesis $H : \omega = \omega_H$ is untrue."

Assume finally that we are interested in the composite hypothesis $H : \omega \in \Omega_H$. Now the large value of the ratio

$$\frac{\max\limits_{\omega \in \Omega} L(\omega; y_{\text{obs}})}{\max\limits_{\omega \in \Omega_H} L(\omega; y_{\text{obs}})}$$

is considered to support the statement "Hypothesis $H : \omega \in \Omega_H$ is untrue."

In all three cases the ratio that is used as a support for an appropriate statement can be expressed in the following form:

$$\frac{L(\hat{\omega}; y_{\text{obs}})}{L(\tilde{\omega}_H; y_{\text{obs}})},$$

where $\hat{\omega}$ is the maximum point of the likelihood function in the *full model* $\mathcal{M}$ and $\tilde{\omega}_H$ is the maximum point of the likelihood function in the model $\mathcal{M}_H$ restricted by the hypothesis. The number $L(\hat{\omega}; y_{\text{obs}})$ is the likelihood under the full model and, respectively, the number $L(\tilde{\omega}_H; y_{\text{obs}})$ is the likelihood under the restricted model.

Example 1.19 Bacterial concentrations data

Let us consider the statistical hypothesis $H : \mu_1 = \mu_2 = \mu_3 = \mu_4 = \mu_5$. The statistical model $\mathcal{S}$ obtained from the statistical model $\mathcal{M}$ by constraining the means to be equal is

$$\mathcal{S} = \text{Submodel}[\mathcal{M}, \theta, \{\theta, \theta, \theta, \theta, \theta\}].$$

The likelihood of the full model $\mathcal{M}$ is equal to 4.62125×10^{-6} and that of the fitted restricted model $\mathcal{S}$ is 1.34938×10^{-7} and so the logarithm of the ratio is 34.2473.   ⬜

1.6.3 General likelihood ratio test statistic (LRT statistic)

The following function of the response

$$w_H(y) = 2(\ell(\hat{\omega}; y) - \ell(\tilde{\omega}_H; y)) = 2 \ln \left(\frac{L(\hat{\omega}; y_{obs})}{L(\tilde{\omega}_H; y_{obs})} \right) \qquad (1.26)$$

is called the *likelihood ratio statistic* of the hypothesis H $: \omega \in \Omega_H$.

Example 1.20 Fuel consumption data (cont'd)

Consider as an example the fuel consumption data and assume that fuel consumption can be modeled as a random sample from some normal distribution whose mean and standard deviation are both unknown and that our interest is in the mean of the distribution. Consider the hypothesis H $: \mu = \mu_H$, where μ_H is some given real number, for example 10.

In Section 1.5.1 it was shown that the log-likelihood function of the mean has the form

$$\begin{aligned}
\ell_\mu(\mu_H; y) &= \max_{0 < \sigma < \infty} \ell(\mu_H, \sigma; y) \\
&= \ell(\mu_H, \tilde{\sigma}_H; y) \\
&= -\frac{n}{2} \ln \left(\frac{\sum_{j=1}^n (y_j - \mu_H)^2}{n} \right) - \frac{n}{2}
\end{aligned}$$

and that the maximum value of the log-likelihood function is

$$\ell(\hat{\mu}, \hat{\sigma}; y) = -\frac{n}{2} \ln \left(\frac{\sum_{j=1}^n (y_j - \bar{y})^2}{n} \right) - \frac{n}{2}.$$

Thus, the likelihood ratio statistic of the hypothesis H $: \mu = \mu_H$ is

$$\begin{aligned}
w_H(y) &= 2(\ell(\hat{\mu}, \hat{\sigma}; y) - \ell(\mu_H, \tilde{\sigma}_H; y)) \\
&= n \ln \left(1 + \frac{1}{n-1} T_H(y)^2 \right),
\end{aligned}$$

where T_H is t-statistics

$$T_H(y) = \frac{\sqrt{n}(\bar{y} - \mu_H)}{\sqrt{\frac{1}{n-1} \sum_{j=1}^n (y_j - \bar{y})^2}} = \frac{\bar{y} - \mu_H}{\frac{s}{\sqrt{n}}}$$

with $n - 1$ degrees of freedom. ⬛

1.6.4   Likelihood ratio test and its observed significance level

The *likelihood ratio test* of the hypothesis $H : \omega \in \Omega_H$ consists of the statement "Hypothesis $H : \omega \in \Omega_H$ is untrue" and of the uncertainty of the statement. The uncertainty of the statement is given by the expression

$$\Pr(\{y \in \mathcal{Y} : w_H(y) \geq w_H(y_{obs})\}; \omega),$$

where the parameter ω satisfies $\omega \in \Omega_H$, provided that the probability is the same or at least approximately the same for all parameters satisfying the condition $\omega \in \Omega_H$. The number describing the uncertainty is called the *observed significance level* of the likelihood ratio test and is often denoted by p_{obs}.

Because every statistical hypothesis $H : \omega \in \Omega_H$ can be expressed in the form $H : g(\omega) = \psi$ for some q-dimensional function g and known vector ψ, the likelihood ratio statistic is equal to

$$w_H(y) = -2r_g(\psi; y).$$

It has been shown that under mild assumptions on the statistical model and function g the random variable $-2r_g(\psi; y)$ has approximately the χ^2-distribution with q degrees of freedom (Cox and Hinkley 1974; Barndorff-Nielsen and Cox 1989, 1994). As the next example shows, in linear models for linear hypotheses on regression coefficients the transformed likelihood ratio statistic

$$F_H(y) = \frac{(n-q)(e^{w_H(y)/n} - 1)}{q}$$

has exactly F-distribution with q nominator and $n - d$ denominator degrees of freedom.

Example 1.21   Fuel consumption data (cont'd)

Consider as an example the fuel consumption data and assume that fuel consummations can be modeled as observations from a linear regression model with temperature and wind velocity as explaining factors. This is a special case of GLM that will be discussed in Chapter 3.

Let us consider the hypothesis $H : \beta = \beta_H$. The log-likelihood function of the model has the form

$$\ell(\beta_1, \beta_2, \beta_3, \sigma; y) = \sum_{j=1}^{n} \left(-\ln(\sigma) - \frac{(y_j - \beta_1 x_{1j} - \beta_2 x_{2j} - \beta_3 x_{3j})^2}{2\sigma^2} \right)$$

$$= -n\ln(\sigma) - \frac{(y - X\beta)^{\mathrm{T}}(y - X\beta)}{2\sigma^2}.$$

At the given values β_H of the regression parameters, the likelihood function is maximized at the following value of the standard deviation:

$$\tilde{\sigma}_H = \sqrt{\frac{(y - X\beta_H)^T(y - X\beta_H)}{n}},$$

and thus the log-likelihood function of β has the form

$$\ell_\beta(\beta_H; y) = \max_{0 < \sigma < \infty} \ell(\beta_H, \sigma; y) = -\frac{n}{2}\ln\left(\frac{(y - X\beta_H)^T(y - X\beta_H)}{n}\right) - \frac{n}{2}.$$

As is well known, the maximum value of the log-likelihood function is

$$\ell(\hat{\beta}_1, \hat{\beta}_2, \hat{\beta}_3, \hat{\sigma}; y) = -\frac{n}{2}\ln\left(\frac{(y - X\hat{\beta})^T(y - X\hat{\beta})}{n}\right) - \frac{n}{2},$$

where $\hat{\beta}$ is the least squares estimate of the regression coefficients

$$\hat{\beta} = (X^T X)^{-1} X^T y.$$

The likelihood ratio statistic of the hypothesis $H : \beta = \beta_H$ has the following form:

$$
\begin{aligned}
w_H(y) &= 2(\ell(\hat{\beta}, \hat{\sigma}; y) - \ell(\beta_H, \tilde{\sigma}_H; y)) \\
&= n\ln\left(\frac{(y - X\beta_H)^T(y - X\beta_H)}{(y - X\hat{\beta})^T(y - X\hat{\beta})}\right) \\
&= n\ln\left(1 + \frac{(\hat{\beta} - \beta_H)^T X^T X(\hat{\beta} - \beta_H)}{(y - X\hat{\beta})^T(y - X\hat{\beta})}\right) \\
&= n\ln\left(1 + \frac{d}{n - d}F_H(y)\right),
\end{aligned}
$$

where $F_H(y)$ is F-statistic

$$F_H(y) = \frac{(\hat{\beta} - \beta_H)^T X^T X(\hat{\beta} - \beta_H)/d}{(y - X\hat{\beta})^T(y - X\hat{\beta})/(n - d)} = \frac{SS_H/d}{SS_E/(n - d)} = \frac{SS_H}{ds^2}.$$

In a similar way as is shown in Section 3.3 of Chapter 3, for a given q-rank $d \times q$ matrix A the likelihood ratio statistic of the hypothesis $H : A^T\beta = \psi$ is

$$w_H(y) = n\ln\left(1 + \frac{q}{n - d}F_H(y)\right),$$

where $F_H(y)$ is F-statistic

$$
\begin{aligned}
F_H(y) &= \frac{(A^T\hat{\beta} - \psi)^T(A^T(X^T X)^{-1}A)^{-1}(A^T\hat{\beta} - \psi)/q}{(y - X\hat{\beta})^T(y - X\hat{\beta})/(n - d)} \\
&= \frac{SS_H/q}{SS_E/(n - d)} = \frac{SS_H}{qs^2}.
\end{aligned}
$$

Because the likelihood ratio statistic of the hypothesis H : $A^{\mathrm{T}}\beta = \psi$ is a monotone function of the F-statistic, the observed significance level for the hypothesis H : $A^{\mathrm{T}}\beta = \psi$ calculated from the observation y_{obs} is equal to

$$\Pr(\{y : w_{\mathrm{H}}(y) \geq w_{\mathrm{H}}(y_{\mathrm{obs}})\}; \beta, \sigma) = \Pr(\{y : F_{\mathrm{H}}(y) \geq F_{\mathrm{H}}(y_{\mathrm{obs}})\}; \beta, \sigma)$$

$$= 1 - F_{\mathrm{F}[q,n-d]\mathrm{distribution}}(F_{\mathrm{H}}(y_{\mathrm{obs}})).$$

When the hypothesis H : $A^{\mathrm{T}}\beta = \psi$ is satisfied the distribution of the F-statistic $F_{\mathrm{H}}(y)$ is F-distribution with q numerator and $n - d$ denominator degrees of freedom.

The model matrix X is

$$\begin{pmatrix} 1 & -3.0 & 15.3 \\ 1 & -1.8 & 16.4 \\ 1 & -10.0 & 41.2 \\ 1 & 0.7 & 9.7 \\ 1 & -5.1 & 19.3 \\ 1 & -6.3 & 11.4 \\ 1 & -15.5 & 5.9 \\ 1 & -4.2 & 24.3 \\ 1 & -8.8 & 14.7 \\ 1 & -2.3 & 16.1 \end{pmatrix}$$

and the statistical model $\mathcal{M}$ is

$$\mathcal{M} = \mathrm{LinearRegressionModel}[X].$$

Consider the hypothesis

$$\begin{pmatrix} 0 & 1 & 0 \\ 0 & 0 & 1 \end{pmatrix} \beta = \begin{pmatrix} 0 \\ 0 \end{pmatrix},$$

that is, $\beta_2 = 0$ and $\beta_3 = 0$. The value of the F-ratio statistic is 32.9385 with observed significance level 0.000275. ☐

Example 1.22 Leukemia data (cont'd)

Assume that the leukemia data can be modeled as statistically independent samples from two Weibull distributions, that is,

$$\mathcal{M} = \mathrm{IndependenceModel}\left[\{\mathrm{SamplingModel}\left[\mathrm{WeibullModel}\left[\kappa_1, \lambda_1\right], 21\right],\right.$$
$$\left.\mathrm{SamplingModel}\left[\mathrm{WeibullModel}\left[\kappa_2, \lambda_2\right], 21\right]\}\right].$$

Table 1.8 contains estimates and approximate 0.95-level profile-likelihood-based confidence intervals of model parameters. Consider the statistical hypothesis H : $\kappa_1 = \kappa_2 = 1$, that is, both distributions are exponential distributions. Observed significance level of this hypothesis is 0.143. ☐

TABLE 1.8: Leukemia data: Estimates and approximate 0.95-level confidence intervals of parameters.

Parameter	Estimate	Lower limit	Upper limit
κ_1	1.354	0.724	2.204
λ_2	33.765	21.848	76.864
κ_2	1.370	0.947	1.880
λ_2	9.482	6.638	13.271

1.7 Maximum likelihood estimate

A distinction without a difference has been introduced by certain writers who distinguish "Point estimation," meaning some process of arriving at an estimate without regard to its precision, from "Interval estimation" in which the precision of the estimate is to some extent taken into account. "Point estimation" in this sense has never been practiced either by myself, or by my predecessor Karl Pearson, who did consider the problem of estimation in some of its aspects, or by his predecessor Gauss of nearly one hundred years earlier, who laid the foundations of the subject (Fisher 1956, p. 143).

Fisher's view on point estimation is taken seriously in this book. Hence, point estimates are never supplied as a means of statistical inference. The concept of the maximum likelihood estimate is, however, practically and technically important because (i) the definition of the relative likelihood function depends on it and (ii) the justification of the asymptotic distribution of the likelihood ratio statistic is usually based on asymptotic normality of the maximum likelihood estimate.

1.7.1 Maximum likelihood estimate (MLE)

If for a given statistical evidence $(y_{\text{obs}}, \mathcal{M})$ with $\mathcal{M} = \{p(y; \omega) : y \in \mathcal{Y}, \omega \in \Omega\}$ there exists a point $\hat{\omega}$ in the parameter space Ω, which maximizes the likelihood function or equivalently the log-likelihood function, that is, if $\hat{\omega}$ satisfies the equation

$$\max\{\ell(\omega; y_{\text{obs}}) : \omega \epsilon \Omega\} = \ell(\hat{\omega}; y_{\text{obs}}), \tag{1.27}$$

the point $\hat{\omega}$ is according to the law of likelihood best supported by the statistical evidence and is called the *maximum likelihood estimate* of ω.

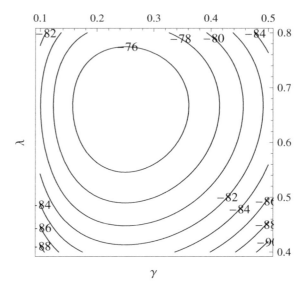

FIGURE 1.20: Respiratory disorders data: Contour plot of the log-likelihood function.

Example 1.23 Respiratory disorders data (cont'd)

Log-likelihood function has the form

$$16\ln(\gamma) + 48\ln(1-\gamma) + 40\ln(\lambda) + 20\ln(1-\lambda).$$

Figure 1.20 shows its contour plot. The maximum likelihood estimate is $(1/4, 2/3)$. Figure 1.21 gives a plot of iteration. ⬜

Example 1.24 Leukemia data (cont'd)

As an example consider the *Control* group of the leukemia data. The statistical model is denoted by

$$\mathcal{M} = \text{SamplingModel}[\text{WeibullModel}[\kappa, \lambda], 21].$$

The log-likelihood function of the *Control* group is

$$-\left(2 + 2^{\kappa+1} + 2^{2\kappa+1} + 2^{3\kappa+2} + 3^{\kappa} + 2^{2\kappa+1}3^{\kappa} + 25^{\kappa} + 211^{\kappa} + 15^{\kappa} + 17^{\kappa} + \right.$$
$$\left. 22^{\kappa} + 23^{\kappa}\right)\left(\frac{1}{\lambda}\right)^{\kappa} + (\kappa - 1)\ln(44201778610176000) + 21\ln(\kappa) - 21\kappa\ln(\lambda).$$

The contour plot is shown in Figure 1.22. This plot seems to indicate that likelihood function has a maximum value near $(1.4, 10)$. The maximum likelihood estimate is $(1.370, 9.482)$. Figure 1.23 gives the steps of iteration. ⬜

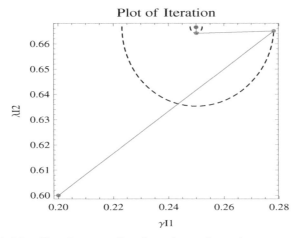

FIGURE 1.21: Respiratory disorders data: Iteration steps of the maximum likelihood estimate.

1.7.2 Asymptotic distribution of MLE

Under very general assumptions, the maximum likelihood estimate has asymptotically normal distribution with ω as mean and inverse of the expected information matrix as variance matrix (Cox and Hinkley 1974).

1.8 Model selection

In previous sections the considered statistical inference procedures were based on the assumption of a given statistical model; that is, the properties of the procedures are valid on the assumption that observations have been generated by some probability model included in the statistical model. They are not suitable to be used as methods for selection of the model.

In the following model selection is done using *Akaike's information criterion* (AIC). The criterion has the form

$$\text{AIC}\left(\mathcal{M}; y_{\text{obs}}\right) = 2\left\{-\ell_{\mathcal{M}}\left(\hat{\omega}; y_{\text{obs}}\right) + d\right\}, \tag{1.28}$$

where d is the dimension of the parameter space Ω of the statistical model $\mathcal{M}$. Among the statistical models under consideration, the models having the smallest value of AIC are preferred. While using Equation (1.28) it is important that the log-likelihood function appearing in it has the "complete" form, that is, includes also those parts that do not depend on the parameter vector. Without using the complete form of the log-likelihood function, the

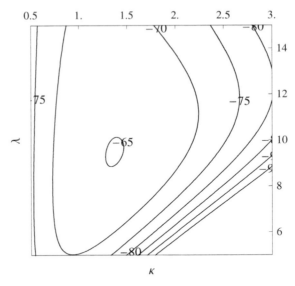

FIGURE 1.22:   Leukemia data: Contour plot of the log-likelihood function.

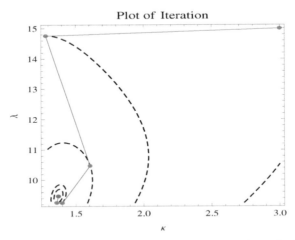

FIGURE 1.23:   Leukemia data: Iteration steps in the calculation of the maximum likelihood estimate.

comparison of statistical models with model functions of different forms is not valid.

Example 1.25 Fuel consumption data (cont'd)

Assume that in the fuel consumption data the amounts of fuel used can be modeled as observations from some linear model in which both, only one, or none of the temperature and wind velocity are used to explain the amount of fuel.

AICs for model with both temperature and velocity as regressors, temperature, velocity, and neither of temperature nor velocity as regressors are 36.90, 43.52, 56.35, and 56.33, respectively. We see that alone wind velocity is not a good predictor, but temperature is. The best model, however, contains both temperature and wind velocity.   ⧠

1.9   Bibliographic notes

Barndorff-Nielsen (1988), Cox (1990), and Lindsey (1995b) are books and articles on modeling. Elementary books on likelihood include Edwards (1972), Kalbfleisch (1985), Azzalini (1996), and Tanner (1996). Pawitan (2001) and Davison (2003) are intermediate books between elementary and advanced. Among advanced books are Fisher (1956), Cox and Hinkley (1974), Barndorff-Nielsen and Cox (1994), Lindsey (1996), Pace and Salvan (1997), and Severini (2000).

Kass in Reid (1988) and O'Hagan (1994) emphasize real-valued interest functions and Clarke (1987), Venzon and Moolgavkar (1988), Chen and Jennrich (1996), and Uusipaikka (1996) discuss calculation of profile likelihood-based confidence intervals.

The basic article of asymptotic distribution of likelihood ratio statistic is Wilks (1938). Other articles and books include Cox (1988), Barndorff-Nielsen and Cox (1989), and Barndorff-Nielsen and Cox (1994). Kullback and Leibler (1951) is the article that introduced Kullback-Leibler divergence and Akaike (1973) introduced AIC. Linhart and Zucchini (1986) is a book on model selection.

General books on probability and statistical inference are Fraser (1968, 1979), Silvey (1980), Grimmet and Welsh (1986), Stirzaker (1994), Welsh (1996), and Knight (2000). Important articles and books on the principles of statistical inference are Birnbaum (1962), Hacking (1965), Durbin (1970), and Berger and Wolpert (1984).

An historical article containing the idea of maximum likelihood is Bernoulli (1777). Articles from Fisher (1912, 1921, 1922, 1925) are both historical and still worth reading. The book from Stigler (1986) is an excellent account of the history of statistical inference before the year 1900.

Chapter 2

Generalized Regression Model

In this chapter first in Section 2.1 examples of various regression situations are given covering a wide range of distributions for responses and relations between regressors and parameters of distributions used to model responses. Then in Section 2.2 the concept of generalized regression model is introduced. In Section 2.3 many well-known regression models are discussed as special cases of generalized regression models. Sections 2.4 and 2.5 contain information on likelihood inference and maximum likelihood estimation, respectively. Finally in Section 2.6 a brief introduction to model checking is presented.

2.1 Examples of regression data

Example 2.1 **Fuel consumption data (cont'd)**

In this case the aim is to explain fuel consumption with help of air temperature and wind velocity. Thus, fuel consumption is called *response variable* and air temperature and wind velocity *regressors* or *explanatory variables*. As fuel consumption is response variable its values are assumed to contain uncertainties described by some distribution, parameters of which depend on regressors. One common assumption is to regard components of response statistically independent observations from some normal distributions with a common standard deviation and with means that depend on regressors. So the model for i^{th} component of the observed response vector is

$$y_i \sim \text{NormalModel}[\mu_i, \sigma]$$

with μ_i being some function of corresponding values of regressors. One simple possibility is to assume that

$$\mu_i = \beta_1 + \beta_2 x_{1i} + \beta_3 x_{2i},$$

where x_{1i} and x_{2i} denote the corresponding values of the regressors and β_1, β_2, and β_3 are unknown *regression parameters*. These assumptions imply that the observed responses can be written in the form

$$y_i = \beta_1 + \beta_2 x_{1i} + \beta_3 x_{2i} + \varepsilon_i$$

with ε_is being modeled as statistically independent and

$$\varepsilon_i \sim \text{NormalModel}[0, \sigma].$$

Thus, the ε_is are modeled as a sample from some normal distribution with zero mean and unknown standard deviation σ.

The statistical model for data can be written in the form

$$y_i \sim \mathcal{M}_i \ \forall i = 1, \ldots, n,$$

where the y_is are statistically independent,

$$\mathcal{M}_i = \{p(y_i; \omega_i) : \ y_i \in \mathcal{Y}_i, \ \omega_i \in \Omega_i\}$$

with $\mathcal{Y}_i = \mathbb{R}$, $\omega_i = (\mu_i, \sigma_i) \in \Omega_i = \mathbb{R} \times (0, \infty)$,

$$\mu_i = \beta_1 + \beta_2 x_{1i} + \beta_3 x_{2i},$$

$$\sigma_1 = \ldots = \sigma_n = \sigma,$$

and

$$p(y_i; \omega_i) = \frac{e^{-\frac{(y_i - \mu_i)^2}{2\sigma^2}}}{\sqrt{2\pi}\sigma}.$$

In compact form

$$y \sim \mathcal{M} = \{p(y; \omega) : \ y \in \mathcal{Y}, \ \omega \in \Omega\},$$

with $\mathcal{Y} = \mathbb{R}^n$, $\omega = (\beta_1, \beta_2, \beta_3, \sigma) \in \Omega = \mathbb{R}^3 \times (0, \infty)$,

$$p(y; \omega) = \prod_{i=1}^{n} \frac{e^{-\frac{(y_i - \mu_i)^2}{2\sigma^2}}}{\sqrt{2\pi}\sigma},$$

and

$$\mu_i = \beta_1 + \beta_2 x_{1i} + \beta_3 x_{2i} \ \forall i = 1, \ldots, n.$$

The important feature is that components of the response vector are statistically independent with their own distributions, which in addition to known values of air temperature and wind velocity also depend on a finite vector $(\beta_1, \beta_2, \beta_3, \sigma)$ of common unknown parameters.

This model is a special case of the *general linear model* (GLM), which will be the subject of Chapter 3. □

TABLE 2.1: Leaf springs data.

Oil	Transfer	Heating	Furnace	Hold down	H1	H2	H3
-	-	-	-	-	7.78	7.78	7.81
-	-	-	+	+	8.15	8.18	7.88
-	-	+	-	+	7.50	7.56	7.50
-	-	+	+	-	7.59	7.56	7.75
-	+	-	-	+	7.94	8.00	7.88
-	+	-	+	-	7.69	8.09	8.06
-	+	+	-	-	7.56	7.62	7.44
-	+	+	+	+	7.56	7.81	7.69
+	-	-	-	-	7.50	7.25	7.12
+	-	-	+	+	7.88	7.88	7.44
+	-	+	-	+	7.50	7.56	7.50
+	-	+	+	-	7.63	7.75	7.56
+	+	-	-	+	7.32	7.44	7.44
+	+	-	+	-	7.56	7.69	7.62
+	+	+	-	-	7.18	7.18	7.25
+	+	+	+	+	7.81	7.50	7.59

Example 2.2 Leaf springs data

Table 2.1 contains free height measurements of leaf springs in the unloaded condition with 8 inches as the target value. The measurements have been done under low and high values of five treatments with three repeats (Pignatiello and Ramberg 1985). The columns of data matrix contain values of oil temperature (-/+), transfer (-/+), heating (-/+), furnace (-/+), hold down (-/+), and three height measurements. Figure 2.1 and Figure 2.2 contain plots of sample means and sample standard deviations against all factors.

In this case the aim is to explain values of response consisting of free height measurements with the help of regressors consisting of oil temperature, transfer, heating, furnace, and hold down. More exactly, free height was measured three times for every observed combination of the regressors for the purpose of finding out how the regressors affect level and variation of free height. One possibility is to assume that for every combination of the regressors the three measurements of free height are a sample from some normal distribution such that both mean and standard deviation of that normal distribution depend on the corresponding values of regressors. Also, all 16 samples are assumed to be statistically independent. Denoting by y_{ij} the jth free height measurement at the ith combination of regressors for $i = 1, \ldots, 16$ and $j = 1, 2, 3$ we have

$$y_{ij} \sim \text{NormalModel}[\mu_i, \sigma_i]$$

with μ_i and σ_i being some functions of corresponding values of regressors. One simple possibility is to assume that means depend on a subset $x_1, \ldots, x_{d-1}$

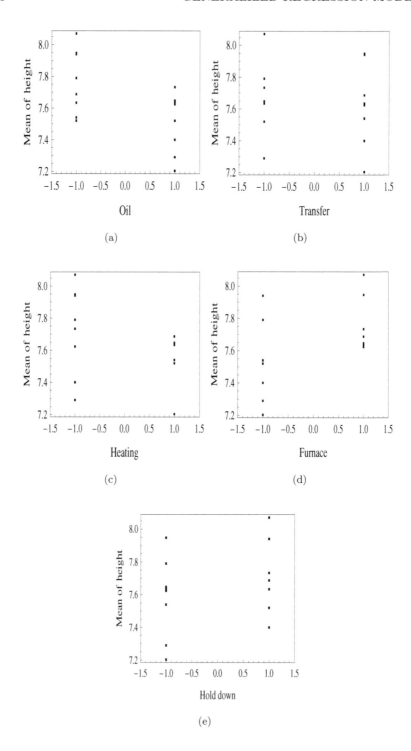

FIGURE 2.1: Leaf springs data: Plots of sample means of three height measurements against the following factors: (a) oil temperature, (b) transfer, (c) heating, (d) furnace, and (e) hold down.

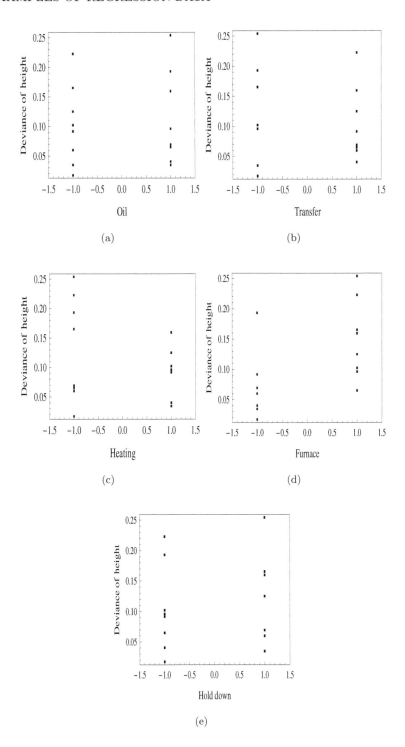

FIGURE 2.2:   Leaf springs data: Plots of sample standard deviations of three height measurements against the following factors: (a) oil temperature, (b) transfer, (c) heating, (d) furnace, and (e) hold down.

and standard deviations on a subset $z_1, \ldots, z_{q-1}$ of regressors so that

$$\mu_i = \beta_1 + \beta_2 x_{1i} + \beta_3 x_{2i} + \cdots + \beta_d x_{d-1,i}$$

and

$$\sigma_i = e^{\gamma_1 + \gamma_2 z_{1i} + \gamma_3 z_{2i} + \cdots + \gamma_q z_{q-1,i}},$$

where x_{ki} and z_{li} denote the corresponding values of the regressors coded so that plus sign $(+)$ is replaced by $+1$ and minus sign $(-)$ by -1. Regression parameters for level, that is, mean, consist of β_1, β_2, ..., and β_d and those for the variation, that is, standard deviation, consist of γ_1, γ_2, ..., and γ_q.

The statistical model for observations can be written in the form

$$y_{ij} \sim \mathcal{M}_{ij} \ \forall i = 1, \ldots, n, \ \forall j = 1, 2, 3,$$

where the y_{ij}s are statistically independent,

$$\mathcal{M}_{ij} = \{p(y_{ij}; \omega_{ij}): \ y_{ij} \in \mathcal{Y}_{ij}, \ \omega_{ij} \in \Omega_{ij}\}$$

with $\mathcal{Y}_{ij} = \mathbb{R}$, $\omega_{ij} = (\mu_{ij}, \sigma_{ij}) \in \Omega_{ij} = \mathbb{R} \times (0, \infty)$,

$$\mu_{ij} = \beta_1 + \beta_2 x_{1i} + \beta_3 x_{2i} + \cdots + \beta_d x_{d-1,i},$$

$$\sigma_{ij} = e^{\gamma_1 + \gamma_2 z_{1i} + \gamma_3 z_{2i} + \cdots + \gamma_q z_{q-1,i}},$$

and

$$p(y_{ij}; \omega_{ij}) = \frac{e^{-\frac{(y_{ij} - \mu_{ij})^2}{2\sigma_{ij}^2}}}{\sqrt{2\pi}\sigma_{ij}}.$$

In compact form

$$y \sim \mathcal{M} = \{p(y; \omega): \ y \in \mathcal{Y}, \ \omega \in \Omega\},$$

with $\mathcal{Y} = \mathbb{R}^{3n}$, $\omega = (\beta_1, \beta_2, \ldots, \beta_d, \gamma_1, \ldots, \gamma_q) \in \Omega = \mathbb{R}^{d+q}$,

$$p(y; \omega) = \prod_{i=1}^{n} \prod_{j=1}^{3} \frac{e^{-\frac{(y_{ij} - \mu_{ij})^2}{2\sigma_{ij}^2}}}{\sqrt{2\pi}\sigma_{ij}}$$

and

$$\mu_{ij} = \beta_1 + \beta_2 x_{1i} + \beta_3 x_{2i} + \cdots + \beta_d x_{d-1,i} \ \forall i = 1, \ldots, n, \ \forall j = 1, 2, 3,$$

$$\sigma_{ij} = e^{\gamma_1 + \gamma_2 z_{1i} + \gamma_3 z_{2i} + \cdots + \gamma_q z_{q-1,i}} \ \forall i = 1, \ldots, n \ \forall j = 1, 2, 3.$$

The important feature is that components of the response vector are statistically independent with their own distributions, which in addition to known values of oil temperature, transfer, heating, furnace, and hold down also depend on a finite vector $(\beta_1, \ldots, \beta_d, \gamma_1, \ldots, \gamma_q)$ of common unknown parameters.

of the response y and contains those components of ω_i that do not depend on i. Also without restriction we can assume that there exists a sequence $z_1, z_2, \ldots, z_n$ of known real vectors and a known function f such that

$$\eta_i = \eta_i(\beta) = f(z_i; \beta) \ \forall i = 1, \ldots, n \tag{2.4}$$

for some unknown d-dimensional vector β. Thus, the GRM can be written in the following form

$$y_i \sim p_i(y_i; f(z_i; \beta), \phi) \ \forall i = 1, \ldots, n \tag{2.5}$$

with $\gamma = (\beta, \phi)$ as the parameter vector of GRM. The sample space of GRM is just the Cartesian product of the sample spaces of the components of the response and the parameter space is a subset of all those real vectors (β, ϕ) for which all the vectors

$$\omega_i = (f(z_i; \beta), \phi) \tag{2.6}$$

belong to the corresponding parameter spaces of model functions p_i. The function f is called the *regression function* and the vector β the *regression parameter vector* of the GRM. GRM is *univariate*, if $m_1 = m_2 = \cdots = m_n = 1$, and *multivariate* otherwise. GRM is a *one-parameter* GRM, if $q_1 = q_2 = \cdots = q_n = 1$, and a *many parameter* GRM otherwise.

To be honest the above considerations introduce just notation, because any parametric statistical model can be expressed in this form. To be a proper GRM the function f and the vectors $z_1, z_2, \ldots, z_n$ must be such that at least two of the components of the response actually depend on the same component of the regression parameter vector β. Informally we can say that a statistical model is a (proper) GRM if it consists of statistically independent parts such that at least two of those parts depend on the same component of the parameter vector of the model.

2.2.4 Regressors and model matrix (matrices)

The components of z_i are called the *regressors* or *regression variables* of the GRM. An important special case arises when for a one-parameter (r-parameter) GRM there exists a real vector valued function g (functions $g_1, \ldots,$ and g_r) such that

$$\eta_i = \eta_i(\beta) = g(z_i)\beta = x_i\beta \ \forall i = 1, \ldots, n, \tag{2.7}$$

where we have denoted by $x_i = g(z_i)$ the known d-dimensional row vectors. In many parameter cases the corresponding equations are

$$\eta_{ij} = \eta_{ij}(\beta) = g_j(z_i)\beta = x_{ij}\beta \ \forall i = 1, \ldots, n, j = 1, \ldots, r. \tag{2.8}$$

Now the components of the regression parameter are called *regression coefficients* and the known $n \times d$ real matrix

$$X = \begin{pmatrix} x_1 \\ x_2 \\ \vdots \\ x_n \end{pmatrix} \tag{2.9}$$

is called the *model matrix* of GRM. In the many parameter case the known $n \times d$ real matrices

$$X_j = \begin{pmatrix} x_{1j} \\ x_{2j} \\ \vdots \\ x_{nj} \end{pmatrix} \quad \forall j = 1, \ldots, r \tag{2.10}$$

are called the *model matrices* of corresponding components of the parameters w_i. In this special case the columns of the model matrix (matrices) are also often called regressors or regression variables.

2.2.5   Example

For the sake of illustration let us assume that in the respiratory disorders data in the *Placebo* group the number of favorable cases was fixed at 16 and in the *Test* group the number of trials was fixed at 60 in the study design. Then it is natural to assume that the number of unfavorable cases has negative binomial distribution in the *Placebo* group and the number of favorable cases has binomial distribution in the *Test* group, respectively.

Now the response $y = (y_1, y_2)$ consists of the number of unfavorable cases in the *Placebo* group and favorable cases in the *Test* group and the two components are statistically independent such that the model function of the first component is

$$p_1(y_1; w_1) = \binom{15 + y_1}{y_1} w_1^{16}(1 - w_1)^{y_1}$$

and that of the second component is

$$p_2(y_2; w_2) = \binom{60}{y_2} w_2^{y_2}(1 - w_2)^{60 - y_2}.$$

The symbols w_1 and w_2 denote the probability of a favorable case in the *Placebo* group and *Test* group, respectively.

Because the purpose of the study is to compare the success probabilities, let us introduce the functions

$$f_1(\gamma_1, \gamma_2) = \frac{\gamma_1}{1 + \gamma_1}$$

and

$$f_2(\gamma_1, \gamma_2) = \frac{\gamma_1 \gamma_2}{1 + \gamma_1 \gamma_2}.$$

Setting these equal to $\omega_1(=\eta_1)$ and $\omega_2(=\eta_2)$, respectively, and solving for γ_1 and γ_2 we get

$$\gamma_1 = \frac{\omega_1}{1 - \omega_1}$$

and

$$\gamma_2 = \frac{(1 - \omega_1)\omega_2}{\omega_1(1 - \omega_2)},$$

that is, $\gamma_1(= \beta_1)$ denotes the odds of a favorable case in the *Placebo* group and $\gamma_2(= \beta_2)$ denotes the ratio of the odds of a favorable case in the *Test* group to that in the *Placebo* group. Finally, let

$$f(x; \beta) = \frac{\beta_1^{x_1} \beta_2^{x_2}}{1 + \beta_1^{x_1} \beta_2^{x_2}} \quad \forall x = (x_1, x_2) \in \mathbb{R}^2, \beta = (\beta_1, \beta_2) \in \mathbb{R}_+^2.$$

Now we can write

$$y_1 \sim p_1(y_1; f((1, 0); \beta))$$

and

$$y_2 \sim p_2(y_2; f((1, 1); \beta)).$$

Thus, we have expressed the statistical model of the response as a univariate and one parameter GRM with model matrix

$$X = \begin{pmatrix} 1 & 0 \\ 1 & 1 \end{pmatrix}$$

and regression function $f(x; \beta)$ defined above.

2.3 Special cases of GRM

2.3.1 Assumptions on parts of GRM

There exist widely used special cases of GRMs. The main groups of special cases arise when more specific assumptions are made concerning

1. structure of the response,

2. number of components of the original parameter vectors depending on regressors,

3. form of model functions of the components of the response, and

4. form of regression function(s).

The common distinctions on these four criteria are the following:

1. uni- or multivariate components in response,

2. one or many components of the original parameter vectors depending on the regression parameter,

3. the components of the response having some special distribution, especially the normal distribution, and

4. regression function(s) being, possibly after some monotone transformation, linear functions of the components of the regression parameter vector.

2.3.2   Various special GRMs

The following is a list of important special cases of GRMs.

- *General linear model* (GLM) is a univariate and one-parameter GRM with normally distributed responses such that means depend linearly on the regression parameters.

- *Nonlinear regression model* is a univariate and one-parameter GRM with normally distributed responses such that means depend nonlinearly on the regression parameters.

- *Multivariate linear regression model* (MANOVA) is a multivariate and many parameter GRM with multinormally distributed responses such that components of mean vectors depend linearly on the regression parameters.

- *Generalized linear model* (GLIM) is a univariate and one-parameter GRM with normally and non-normally distributed responses such that transformed means depend linearly on regression parameters. Generalized linear models include (at least) binomial, Poisson, normal, inverse Gaussian, and gamma distributions.

- *Weighted GLM* is a univariate and two-parameter GRM with normally distributed responses such that the means depend linearly on the regression parameters and the variances are equal to the common scale parameter divided by the known weights.

- *Weighted nonlinear regression model* is a univariate and two-parameter GRM with normally distributed responses such that the means depend nonlinearly on the regression parameters and the variances are equal to the common scale parameter divided by the known weights.

- *Quality design model* or *"Taguchi" model* is a univariate and two-para-meter GRM with normally distributed responses such that the means and the transformed variances depend linearly on the regression param-eters.

- *Multinomial regression model* is a multivariate and many parameter GRM with multinomially distributed responses such that the compo-nents of transformed probability vectors depend linearly on the regres-sion parameters.

- *Lifetime regression model* is a univariate and one-parameter GRM with positively distributed responses usually such that the transformed scale parameters depend linearly on the regression parameters. Common dis-tributions for the components of the response are exponential, gamma, inverse Gaussian, log-normal, and Weibull distributions. In lifetime re-gression models observations are often censored.

- *Cox regression model* is a univariate and one-parameter GRM with pos-itively distributed responses usually such that the transformed hazard ratios depend linearly on the regression parameters. As in lifetime re-gression models, observations are often censored.

2.4   Likelihood inference

In this and the next sections we follow Davison (2003, pp. 471-477) by as-suming that the model is univariate one-parameter GRM with one-dimensional common parameter ϕ, if there exists one. Also assume that the functional forms of the model functions of the components of the response vector are the same $p(y_i; \eta_i, \phi)$. The log-likelihood function of GRM is

$$\ell(\beta, \phi; y_{\text{obs}}) = \sum_{i=1}^{n} \ln(p(y_i; \eta_i(\beta), \phi))$$

$$= \sum_{i=1}^{n} \ell(\eta_i(\beta), \phi; y_i) \tag{2.11}$$

Assume first that ϕ is known. The maximum likelihood estimate of η_i under the model is $\hat{\eta}_i = \eta_i(\hat{\beta})$, where $\hat{\beta}$ is the maximum likelihood estimate of β. In the *saturated model* with no constraints on the η_i the maximum likelihood estimate of η_i is $\tilde{\eta}_i$, which maximizes $\ell(\eta_i; y_i)$. Then the *scaled deviance* is defined as

$$D = 2\sum_{i=1}^{n}\{\ell(\tilde{\eta}_i; y_i) - \ell(\hat{\eta}_i; y_i)\}$$

$$= 2\left\{\sum_{i=1}^{n}\ell(\tilde{\eta}_i; y_i) - \sum_{i=1}^{n}\ell(\hat{\eta}_i; y_i)\right\}, \qquad (2.12)$$

that is, as the likelihood ratio statistic comparing GRM with the saturated model.

Suppose we want to test a hypothesis consisting of q linearly independent restrictions on β. Let A denote the model in which all d are unconstrained and let B denote the model satisfying the q constraints on β, with maximum likelihood estimates $\hat{\eta}^A$ and $\hat{\eta}^B$ and deviancies D_A and D_B, respectively. The the likelihood ratio test statistic for comparing the two models is

$$2\sum_{i=1}^{n}\{\ell(\hat{\eta}_i^A; y_i) - \ell(\hat{\eta}_i^B; y_i)\} = D_A - D_B, \qquad (2.13)$$

and provided the models are regular this has an approximate χ^2-distribution with q degrees of freedom under the model B.

If ϕ is unknown, it is replaced by an estimate. The large-sample properties of deviance differences outlined above still apply, though in small samples it may be better to replace the approximating χ^2-distribution by an F-distribution with numerator degrees of freedom equal to the degrees of freedom for the estimation of ϕ. In Section 3.3 it will be shown that the F-distribution is exact for linear constraints on regression parameters in GLM.

2.5　MLE with iterative reweighted least squares

If ϕ is known, the log-likelihood function can be written in the form

$$\ell(\beta) = \sum_{i=1}^{n}\ell_i(\eta_i(\beta), \phi), \qquad (2.14)$$

where dependence on the values of response has been suppressed and

$$\ell_i(\eta_i(\beta), \phi) = \ln(p(y_i; \eta_i(\beta), \phi))$$

is the contribution made by the ith observation. The maximum likelihood estimate $\hat{\beta}$ usually satisfies the equation

$$\frac{\partial\ell(\hat{\beta})}{\partial\beta_r} = 0 \ \forall r = 1, \ldots, d, \qquad (2.15)$$

which can be written in matrix form as follows

$$\frac{\partial \ell(\hat{\beta})}{\partial \beta} = \frac{\partial \eta^{\mathrm{T}}}{\partial \beta} u(\hat{\beta}) = 0,$$

where the ith element of the $n \times 1$ vector $u(\beta)$ is $\partial \ell(\beta)/\partial \eta_i$.

Taylor series expansion with respect to trial value β gives

$$\frac{\partial \eta^{\mathrm{T}}}{\partial \beta} u(\beta) + \left\{ \sum_{i=1}^{n} \frac{\partial \eta_i(\beta)}{\partial \beta} \frac{\partial^2 \ell_i(\beta)}{\partial \eta_i^2} \frac{\partial \eta_i(\beta)}{\partial \beta^{\mathrm{T}}} + \sum_{i=1}^{n} \frac{\partial^2 \eta_i(\beta)}{\partial \beta \partial \beta^{\mathrm{T}}} u_i(\beta) \right\} (\hat{\beta} - \beta) \doteq 0.$$

Denoting by $-J(\beta)$ the $d \times d$ matrix in braces on the left and assuming it to be invertible we get

$$\hat{\beta} \doteq \beta + J(\beta)^{-1} \frac{\partial \eta^{\mathrm{T}}}{\partial \beta} u(\beta).$$

So the maximum likelihood estimate may be obtained by iteration starting from a particular β. The iteration method is the Newton-Raphson algorithm applied to a special setting. In practice it may be more convenient to replace $J(\beta)$ by its expected value

$$I(\beta) = \sum_{i=1}^{n} \frac{\partial \eta_i(\beta)}{\partial \beta} \mathrm{E} \left(-\frac{\partial^2 \ell_i(\beta)}{\partial \eta_i^2} \right) \frac{\partial \eta_i(\beta)}{\partial \beta^{\mathrm{T}}}$$

with the other term vanishing because $\mathrm{E}\{u_i(\beta)\} = 0$. This may be written in the form

$$I(\beta) = X(\beta)^{\mathrm{T}} W(\beta) X(\beta),$$

where $X(\beta)$ is the $n \times d$ matrix $\partial \eta(\beta)/\partial \beta^{\mathrm{T}}$ and $W(\beta)$ is the diagonal matrix whose i^{th} diagonal element is $\mathrm{E}(-\partial^2 \ell_i(\beta)/\partial \eta_i^2)$.

Replacing $J(\beta)$ by $X(\beta)^{\mathrm{T}} W(\beta) X(\beta)$ and reorganizing we obtain

$$\hat{\beta} = \left(X^{\mathrm{T}} W X \right)^{-1} X^{\mathrm{T}} W \left(X\beta + W^{-1}u \right) = \left(X^{\mathrm{T}} W X \right)^{-1} X^{\mathrm{T}} W z, \qquad (2.16)$$

where dependence on β has been suppressed and $z = X\beta + W^{-1}u$. Thus, the calculation of the next value of iteration can be interpreted as applying weighted linear regression of the vector z on the columns of $X(\beta)$, using the weight matrix $W(\beta)$. The maximum likelihood estimates are obtained by repeating this step until the log-likelihood, the estimates, or more often both are essentially unchanged. The variable z plays the role of the response in the weighted least squares step and is sometimes called the *adjusted response*.

Often the structure of a model simplifies the estimation of an unknown value of ϕ. It may be estimated by a separate step between iterations of $\hat{\beta}$, by including it in the iteration step, or from the profile log-likelihood $\ell_P(\phi)$ of ϕ.

Example 2.9 **General linear model (GLM)**

In GLM we have $\mu_i = \eta_i = x_i\beta$ and the log-likelihood function $\ell_i(\eta_i, \sigma^2)$ has the form

$$\ell_i(\eta_i, \sigma^2) = -\frac{1}{2}\left\{\ln(2\pi\sigma^2) + \frac{1}{\sigma^2}(y_i - \eta_i)^2\right\},$$

so

$$u_i(\eta_i) = \frac{\partial\ell_i}{\partial\eta_i} = \frac{1}{\sigma^2}(y_i - \eta_i), \quad \frac{\partial^2\ell_i}{\partial\eta_i^2} = -\frac{1}{\sigma^2}.$$

That is, the ith element of the diagonal matrix W is the constant $1/\sigma^2$. The (i, r) element of the matrix $\partial\eta/\partial\beta^{\text{T}}$ is $\partial\eta_i/\partial\beta_r = x_{ir}$, and so $X(\beta)$ is simply the $n \times d$ model matrix X. The adjusted response $z = X(\beta)\beta + W(\beta)^{-1}u(\beta) = y$. In this case iterative weighted least squares converges in a single step because it is equivalent to ordinary least squares.

The maximum likelihood estimate of σ^2 is easily obtained from profile likelihood $\ell_P(\sigma^2)$ and it is $\hat\sigma^2 = SS(\hat\beta)/n$, where $SS(\beta)$ is the sum of squares $(y - X\beta)^{\text{T}}(y - X\beta)$.

GLM will be considered in full detail later in Chapter 3.   ⬚

Example 2.10 **Nonlinear regression model**

In the nonlinear regression model we have $\mu_i = \eta_i(\beta)$, which is a proper nonlinear function of β, and the log-likelihood function $\ell_i(\eta_i, \sigma^2)$ has the same form as in the previous example, as have u and W as well. The ith row of the matrix $X = \partial\eta/\partial\beta^{\text{T}}$, however, is $(\partial\eta_i/\partial\beta_1, \ldots, \partial\eta_i/\partial\beta_d)$ and depends on β. It can be shown that the new value for $\hat\beta$ has the form

$$\hat\beta \doteq (X^{\text{T}}X)^{-1}X^{\text{T}}(X\beta + y - \eta),$$

where X and η are evaluated at the current β. Here $\eta \neq X\beta$ and so the maximum likelihood estimate results from iteration.

The log-likelihood function of σ^2 is

$$\ell_P(\sigma^2) = \max_{\beta} \ell(\beta, \sigma^2) = -\frac{1}{2}\left\{n\ln(2\pi\sigma^2) + \frac{SS(\hat\beta)}{\sigma^2}\right\}$$

so the maximum likelihood estimate of σ^2 is $\hat\sigma^2 = SS(\hat\beta)/n$.

The nonlinear regression model will be considered in full detail in Chapter 4.   ⬚

2.6   Model checking

The *deviance residual* of the ith component of the response vector is

$$d_i = \text{sign}(\tilde\eta_i - \hat\eta_i)[2\{\ell_i(\tilde\eta_i, \phi; y_i) - \ell_i(\hat\eta_i, \phi; y_i)\}]^{1/2}. \tag{2.17}$$

The *standardized deviance residuals* have the form

$$r_{Di} = \frac{d_i}{\sqrt{1 - h_{ii}}}, \tag{2.18}$$

where h_{ii} is called the *leverage* of ith component of the response vector and is the ith diagonal element of the matrix

$$H(\hat{\beta}) = W(\hat{\beta})^{1/2} X(\hat{\beta}) (X(\hat{\beta})^{\text{T}} W(\hat{\beta}) X(\hat{\beta}))^{-1} X(\hat{\beta})^{\text{T}} W(\hat{\beta})^{1/2}. \tag{2.19}$$

Standardized residuals are more homogeneous and usually they have roughly unit variances and distributions close to normal, though possibly with non-zero mean. Exceptions to this are binary data and Poisson data with small responses, where residuals are essentially discrete.

A better general definition of residual combines the standardized deviance residual with the *standardized Pearson residual*

$$r_{Pi} = \frac{u_i(\hat{\beta})}{\sqrt{w_i(\hat{\beta})(1 - h_{ii})}}, \tag{2.20}$$

which is a standardized score statistic. Detailed calculations show that the distributions of the quantities

$$r_i^* = r_{Di} + \frac{1}{r_{Di}} \ln\left(\frac{r_{Pi}}{r_{Di}}\right), \tag{2.21}$$

are close to normal for a wide range of models. For a normal linear model, that is, for GLM, $r_{Pi} = r_{Di} = r_i^* = r_i$, where r_i is the usual standardized residual of GLM.

The *influence* of the ith case is measured by

$$2\{\ell(\hat{\beta}) - \ell(\hat{\beta}_{-i})\}/d, \tag{2.22}$$

where $\hat{\beta}_{-i}$ is the estimate of β with the ith case deleted from the model. Calculation of all these influences requires n additional fits, and it is more convenient to use the *approximate Cook statistic*

$$C_i = \frac{h_{ii}}{d(1 - h_{ii})} r_{Pi}^2. \tag{2.23}$$

2.7 Bibliographic notes

The idea of generalized regression models was introduced by Lindsey (1974a,b). Another source containing discussion of generalized regression model is Davison (2003).

The Cook statistic was introduced in Cook (1977, 1979).

Chapter 3

General Linear Model

The main topic of Gauss's book was a masterly investigation of the mathematics of planetary orbits. At the end, however, he added a section on the combination of observations. In that section he addressed essentially the same problem Mayer, Euler, Boscovich, Laplace, and Legendre had addressed, but with one significant difference: Gauss couched the problem in explicitly probabilistic terms. (Gauss 1809; Stigler 1986, pp. 140-141)

In this Chapter in Section 3.1 definition of the general linear model is presented. In Section 3.2 estimates of regression coefficients are derived first using least squares and then the maximum likelihood method. Then in Section 3.3 the likelihood ratio test of linear hypotheses on regression coefficients is considered. In Section 3.4 various profile likelihood-based confidence procedures are discussed. These include joint confidence regions for finite sets of linear combinations of regression coefficients and confidence intervals for linear combinations of regression coefficients. Finally, in Section 3.4 model criticism in terms of residuals is discussed.

3.1 Definition of the general linear model

Gauss supposed that there were unknown linear functions

$$
\begin{aligned}
V &= ap + bq + cr + ds + \text{etc.} \\
V' &= a'p + b'q + c'r + d's + \text{etc.} \\
V'' &= a''p + b''q + c''r + d''s + \text{etc.} \\
&\text{etc.}
\end{aligned}
$$

of observables $a, b, c, \ldots$ with unknown coefficients $p, q, r, \ldots$, and that these are found by direct observation to be M, M', M'', etc. Gauss supposed that the possible values of errors $\Delta = V - M$, $\Delta' = V' - M'$, $\Delta'' = V'' - M''$, etc. had probabilities given by a curve $\phi(\Delta), \phi(\Delta'), \phi(\Delta'')$ etc

$$\phi(\Delta) = \frac{h}{\sqrt{\pi}} e^{-h^2 \Delta^2}$$

for some positive constant h, where h could be viewed as a measure of precision of observation. (Gauss 1809; Stigler 1986, pp. 140-141)

In Section 2.3 the general linear model (GLM) was characterized as a univariate and one-parameter GRM with statistically independent normally distributed components of response vector such that means of the components depend linearly on regression coefficients. Classically this is expressed with the formula

$$y_i = x_i \beta + \varepsilon_i \ \forall i = 1, \ldots, n \tag{3.1}$$

or in matrix form

$$y = X\beta + \varepsilon, \tag{3.2}$$

where y is an n-dimensional response vector and ε is an n-dimensional error vector, whose components are statistically independent and normally distributed with zero means and common unknown standard deviation σ. The matrix X is an $n \times d$ model matrix and β is a d-dimensional vector of unknown regression coefficients. Without restriction one can assume that the model matrix is nonsingular, that is, has column rank d. Also the number of observations must be greater than d because otherwise there would be too few observations to estimate all the parameters of the model.

Equivalently we can say that the response vector y has n-variate normal distribution with mean vector

$$E(y) = X\beta \tag{3.3}$$

and variance matrix

$$\text{var}(y) = \sigma^2 I_n. \tag{3.4}$$

From the assumptions follows that the likelihood function of the model is

$$L(\beta, \sigma; y) = \prod_{i=1}^{n} \frac{1}{\sqrt{2\pi}\sigma} e^{-\frac{(y_i - x_i \beta)^2}{2\sigma^2}}$$

and the log-likelihood function is

$$\ell(\beta, \sigma; y) = -\frac{n}{2} \ln(2\pi) - n \ln(\sigma) - \frac{1}{2\sigma^2} \sum_{i=1}^{n} (y_i - x_i \beta)^2$$

$$= -\frac{n}{2} \ln(2\pi) - n \ln(\sigma) - \frac{1}{2\sigma^2} (y - X\beta)^{\text{T}} (y - X\beta). \tag{3.5}$$

TABLE 3.1:

Tensile strength data.

Hardness	Strength
52	12.3
54	12.8
56	12.5
57	13.6
60	14.5
61	13.5
62	15.6
64	16.1
66	14.7
68	16.1
69	15.0
70	16.0
71	16.7
71	17.4
73	15.9
76	16.8
76	17.6
77	19.0
80	18.6
83	18.9

Example 3.1 Tensile strength data

Data were obtained on the hardness and tensile strengths of 20 samples of an alloy produced under a variety of conditions (Brooks and Dick 1951). The two columns in Table 3.1 contain values of hardness and tensile strength. In Figure 3.1 observations of tensile strength are plotted against corresponding observations of hardness.

Simple linear regression model for tensile strength is a special case of GLM, in which in addition to regression coefficients response depends linearly on the numerical regressor hardness, that is,

$$y_i = \beta_1 + \beta_2 x_i + \varepsilon_i \ \forall i = 1, \ldots, 20,$$

where y_i denotes i^{th} value of tensile strength and x_i the corresponding value of hardness. Thus, the graph of the regression function in $\mathbb{R}^2$ as the function of hardness is a line. Figure 3.1 gives that line in the case of $\beta_1 = 2$ and $\beta_2 = 0.2$.

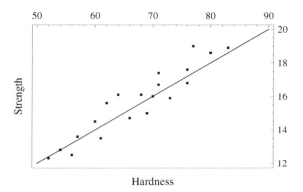

FIGURE 3.1: Tensile strength data: Tensile strength on hardness.

In matrix form we get

$$
\begin{pmatrix} 12.3 \\ 12.8 \\ 12.5 \\ \vdots \\ 19.0 \\ 18.6 \\ 18.9 \end{pmatrix}
=
\begin{pmatrix} 1 & 52 \\ 1 & 54 \\ 1 & 56 \\ \vdots & \vdots \\ 1 & 77 \\ 1 & 80 \\ 1 & 83 \end{pmatrix}
\begin{pmatrix} \beta_1 \\ \beta_2 \end{pmatrix}
+
\begin{pmatrix} \varepsilon_1 \\ \varepsilon_2 \\ \varepsilon_3 \\ \vdots \\ \varepsilon_{18} \\ \varepsilon_{19} \\ \varepsilon_{20} \end{pmatrix},
$$

where the ε_is are modeled as a random sample from the normal distribution with zero mean and unknown standard deviation σ. This statistical model is denoted by

$$\mathcal{M} = \mathrm{LinearRegressionModel}[X],$$

where X denotes the model matrix

$$
X =
\begin{pmatrix} 1 & 52 \\ 1 & 54 \\ 1 & 56 \\ \vdots & \vdots \\ 1 & 77 \\ 1 & 80 \\ 1 & 83 \end{pmatrix}.
$$

□

Example 3.2 Fuel consumption data (cont'd)

The following defines a *multiple linear regression model* for fuel consumption

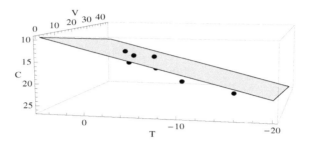

FIGURE 3.2:   Fuel consumption data: Fuel consumption against air temperature and wind velocity with a regression plane.

with air temperature and wind velocity as regressors

$$y_i = \beta_1 + \beta_2 x_{i1} + \beta_2 x_{i2} + \varepsilon_i \ \forall i = 1, \dots, 10,$$

where y_i denotes the ith value of fuel consumption and x_{i1} and x_{i2} the corresponding values of air temperature and wind velocity, respectively. Thus, the graph of the regression function in $\mathbb{R}^3$ as the function of air temperature and wind velocity is a plane. Figure 3.2 shows that plane in the case of $\beta_1 = 12$, $\beta_2 = -0.5$, and $\beta_3 = 0.1$.

In matrix form we get

$$
\begin{pmatrix} 14.96 \\ 14.10 \\ 23.76 \\ \vdots \\ 16.25 \\ 20.98 \\ 16.88 \end{pmatrix}
=
\begin{pmatrix} 1 & -3.0 & 15.3 \\ 1 & -1.8 & 16.4 \\ 1 & -10.0 & 41.2 \\ \vdots & \vdots & \vdots \\ 1 & -4.2 & 24.3 \\ 1 & -8.8 & 14.7 \\ 1 & -2.3 & 16.1 \end{pmatrix}
\begin{pmatrix} \beta_1 \\ \beta_2 \\ \beta_3 \end{pmatrix}
+
\begin{pmatrix} \varepsilon_1 \\ \varepsilon_2 \\ \varepsilon_3 \\ \vdots \\ \varepsilon_8 \\ \varepsilon_9 \\ \varepsilon_{10} \end{pmatrix},
$$

where the ε_is are modeled as a random sample from the normal distribution with zero mean and unknown standard deviation σ. This statistical model is denoted by

$$\mathcal{M} = \text{LinearRegressionModel}[X],$$

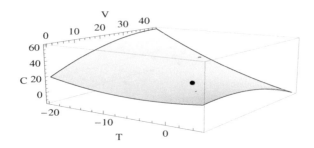

FIGURE 3.3:   Fuel consumption data: Fuel consumption against air temperature and wind velocity with a regression surface.

where X is the model matrix

$$X = \begin{pmatrix} 1 & -3.0 & 15.3 \\ 1 & -1.8 & 16.4 \\ 1 & -10.0 & 41.2 \\ \vdots & \vdots & \vdots \\ 1 & -4.2 & 24.3 \\ 1 & -8.8 & 14.7 \\ 1 & -2.3 & 16.1 \end{pmatrix}.$$

A more complicated linear model is the following multiple linear regression model for fuel consumption where the regression function is quadratic with respect to both air temperature and wind velocity

$$y_i = \beta_1 + \beta_2 x_{i1} + \beta_3 x_{i2} + \beta_4 x_{i1}^2 + \beta_5 x_{i1} x_{i2} + \beta_6 x_{i2}^2 + \varepsilon_i \ \forall i = 1, \ldots, 10,$$

where the y_i denotes i^{th} value of fuel consumption and x_{i1} and x_{i2} the corresponding values of air temperature and wind velocity, respectively. Thus, the graph of the regression function in $\mathbb{R}^3$ as the function of air temperature and wind velocity is a quadratic surface. This kind of model is often said to be a *response surface model*. Figure 3.3 shows that surface in the case of $\beta_1 = 10$, $\beta_2 = 0.5$, $\beta_3 = 0.6$, $\beta_4 = 0.05$, $\beta_5 = -0.06$, and $\beta_6 = -0.02$.

In matrix form we get

$$\begin{pmatrix} 14.96 \\ 14.10 \\ 23.76 \\ \vdots \\ 16.25 \\ 20.98 \\ 16.88 \end{pmatrix} = \begin{pmatrix} 1 & -3.0 & 15.3 & 9.00 & -45.90 & 234.09 \\ 1 & -1.8 & 16.4 & 3.24 & -29.52 & 268.96 \\ 1 & -10.0 & 41.2 & 100.00 & -412.00 & 1697.44 \\ \vdots & \vdots & \vdots & \vdots & \vdots & \vdots \\ 1 & -4.2 & 24.3 & 17.64 & -102.06 & 590.49 \\ 1 & -8.8 & 14.7 & 77.44 & -129.36 & 216.09 \\ 1 & -2.3 & 16.1 & 5.29 & -37.03 & 259.21 \end{pmatrix} \begin{pmatrix} \beta_1 \\ \beta_2 \\ \beta_3 \\ \beta_4 \\ \beta_5 \\ \beta_6 \end{pmatrix} + \begin{pmatrix} \varepsilon_1 \\ \varepsilon_2 \\ \varepsilon_3 \\ \vdots \\ \varepsilon_8 \\ \varepsilon_9 \\ \varepsilon_{10} \end{pmatrix},$$

where the ε_is are modeled as a random sample from the normal distribution with zero mean and unknown standard deviation σ. This statistical model is denoted in the same way as previous ones with the appropriate model matrix.
⬚

3.2 Estimate of regression coefficients

Of all the principles that can be proposed for this purpose, I think there is none more general, more exact, or easier to apply, than that which we have used in this work; it consists of making the sum of the squares of the errors a *minimum*. By this method, a kind of equilibrium is established among the errors which, since it prevents the extremes from dominating, is appropriate for revealing the state of the system which most nearly approaches the truth (Legendre 1805, pp. 72-73).

3.2.1 Least squares estimate (LSE)

The results considered in this section are more general than the assumptions made in Section 3.1 imply. Namely, instead of assuming that the components of the response vector are statistically independent, it suffices to assume that their mean vector is $X\beta$, they have a common variance σ^2, and they are uncorrelated without any assumption being made about their distributional form. If, however, the normal distribution assumption is made the assumption of uncorrelatedness implies statistical independence. One can be even slightly more general and because this more general assumption is needed later on that assumption, namely the assumption of the covariance matrix of the response vector being known up to an unknown scale parameter, is made in this section. Thus, the covariance matrix of the response vector is assumed to have the form

$$\Sigma = \sigma^2 V, \qquad (3.6)$$

where V is a known nonsingular positive definite $n \times n$ matrix.

The *least squares estimate* (LSE) of the location parameter vector β is that value of β, which minimizes the expression

$$\sum_{i=1}^{n} e_i^2 = \sum_{i=1}^{n}(y_i - x_i\beta)^2 = (y - X\beta)^{\mathrm{T}}(y - X\beta) = |y - X\beta|^2 \qquad (3.7)$$

or in the more general case the expression

$$e^{\mathrm{T}}V^{-1}e = (y - X\beta)^{\mathrm{T}}V^{-1}(y - X\beta) \qquad (3.8)$$

where x_i is the ith row of the model matrix. The LSE b will be obtained as the solution of the following Gauss *normal equation*

$$X^\mathrm{T} X \beta = X^\mathrm{T} y \tag{3.9}$$

or more generally

$$X^\mathrm{T} V^{-1} X \beta = X^\mathrm{T} V^{-1} y. \tag{3.10}$$

When the columns of the model matrix are linearly independent, which assumption is always made in this book, one gets

$$b = (X^\mathrm{T} X)^{-1} X^\mathrm{T} y \tag{3.11}$$

and, once again, more generally

$$b = (X^\mathrm{T} V^{-1} X)^{-1} X^\mathrm{T} V^{-1} y. \tag{3.12}$$

This minimizes the sum of squared errors because

$$
\begin{aligned}
(y - X\beta)^\mathrm{T} V^{-1}(y - X\beta) &= (y - Xb + X(b - \beta))^\mathrm{T} V^{-1}(y - Xb + X(b - \beta)) \\
&= (y - Xb)^\mathrm{T} V^{-1}(y - Xb) \\
&\quad + 2(b - \beta)^\mathrm{T} X^\mathrm{T} V^{-1}(y - Xb) \\
&\quad + (X(b - \beta))^\mathrm{T} X(b - \beta).
\end{aligned}
$$

Now as the LSE satisfies the Gauss normal equation the middle term vanishes and so

$$(y - X\beta)^\mathrm{T} V^{-1}(y - X\beta) = (y - Xb)^\mathrm{T} V^{-1}(y - Xb) + |X(b - \beta)|^2.$$

Clearly this depends on β only through the length of the vector $X(b - \beta)$ and so the expression is minimized by setting this vector equal to zero. From the assumption of the linear independence of the columns of the model matrix follows that β must be equal to b at the minimum.

The mean vector of b is

$$
\begin{aligned}
\mathrm{E}(b) &= \mathrm{E}((X^\mathrm{T} V^{-1} X)^{-1} X^\mathrm{T} V^{-1} y) \\
&= (X^\mathrm{T} V^{-1} X)^{-1} X^\mathrm{T} V^{-1} \mathrm{E}(y) \\
&= (X^\mathrm{T} V^{-1} X)^{-1} X^\mathrm{T} V^{-1} X \beta \\
&= \beta
\end{aligned}
\tag{3.13}
$$

and the variance matrix is

$$
\begin{aligned}
\mathrm{var}(b) &= \mathrm{var}((X^\mathrm{T} V^{-1} X)^{-1} X^\mathrm{T} V^{-1} y) \\
&= (X^\mathrm{T} V^{-1} X)^{-1} X^\mathrm{T} V^{-1} \mathrm{var}(y) V^{-1} X (X^\mathrm{T} V^{-1} X)^{-1} \\
&= \sigma^2 (X^\mathrm{T} V^{-1} X)^{-1}.
\end{aligned}
\tag{3.14}
$$

Finally, the LSE and the vector of *residuals* $e = y - Xb$ are uncorrelated because

$$
\begin{aligned}
\operatorname{cov}(b, y - Xb) &= \operatorname{cov}((X^TV^{-1}X)^{-1}X^TV^{-1}y, I - X(X^TV^{-1}X)^{-1}X^TV^{-1}y) \\
&= \sigma^2(X^TV^{-1}X)^{-1}X^TV^{-1}\operatorname{var}(y)(I - V^{-1}X(X^TV^{-1}X)^{-1}X^T) \\
&= 0.
\end{aligned} \tag{3.15}
$$

Under the above-made assumptions c^Tb is the *best linear unbiased estimator* (*BLUE*) of the linear combination $c^T\beta$ of the regression coefficients.

Example 3.3 Tensile strength data (cont'd)

The LSE of the vector of intercept and slope are

$$
\left(\begin{pmatrix} 1 & 52 \\ 1 & 54 \\ 1 & 56 \\ \vdots & \vdots \\ 1 & 77 \\ 1 & 80 \\ 1 & 83 \end{pmatrix}^T \begin{pmatrix} 1 & 52 \\ 1 & 54 \\ 1 & 56 \\ \vdots & \vdots \\ 1 & 77 \\ 1 & 80 \\ 1 & 83 \end{pmatrix}\right)^{-1} \begin{pmatrix} 1 & 52 \\ 1 & 54 \\ 1 & 56 \\ \vdots & \vdots \\ 1 & 77 \\ 1 & 80 \\ 1 & 83 \end{pmatrix}^T \begin{pmatrix} 12.3 \\ 12.8 \\ 12.5 \\ \vdots \\ 19.0 \\ 18.6 \\ 18.9 \end{pmatrix} = \begin{pmatrix} 20 & 1346 \\ 1346 & 92088 \end{pmatrix}^{-1} \begin{pmatrix} 313.6 \\ 21433. \end{pmatrix}
$$

$$
= \begin{pmatrix} 0.99783 \\ 0.21816 \end{pmatrix}.
$$

□

Example 3.4 Fuel consumption data (cont'd)

The least squares estimate of the vector of regression coefficients is

$$
\left(\begin{pmatrix} 1 & -3.0 & 15.3 \\ 1 & -1.8 & 16.4 \\ 1 & -10.0 & 41.2 \\ \vdots & \vdots & \vdots \\ 1 & -4.2 & 24.3 \\ 1 & -8.8 & 14.7 \\ 1 & -2.3 & 16.1 \end{pmatrix}^T \begin{pmatrix} 1 & -3.0 & 15.3 \\ 1 & -1.8 & 16.4 \\ 1 & -10.0 & 41.2 \\ \vdots & \vdots & \vdots \\ 1 & -4.2 & 24.3 \\ 1 & -8.8 & 14.7 \\ 1 & -2.3 & 16.1 \end{pmatrix}\right)^{-1} \begin{pmatrix} 1 & -3.0 & 15.3 \\ 1 & -1.8 & 16.4 \\ 1 & -10.0 & 41.2 \\ \vdots & \vdots & \vdots \\ 1 & -4.2 & 24.3 \\ 1 & -8.8 & 14.7 \\ 1 & -2.3 & 16.1 \end{pmatrix}^T \begin{pmatrix} 14.96 \\ 14.10 \\ 23.76 \\ \vdots \\ 16.25 \\ 20.98 \\ 16.88 \end{pmatrix} =
$$

$$
\begin{pmatrix} 10 & -56.3 & 174.3 \\ -56.3 & 519.05 & -1010.78 \\ 174.3 & -1010.78 & 3897.63 \end{pmatrix}^{-1} \begin{pmatrix} 177.35 \\ -1129.32 \\ 3221.31 \end{pmatrix} = \begin{pmatrix} 11.934 \\ -0.629 \\ 0.130 \end{pmatrix}.
$$

□

3.2.2   Maximum likelihood estimate (MLE)

In a GLM, the *maximum likelihood estimate* (MLE) $\hat{\beta}$ of the vector of regression coefficients β is the same as the LSE b and its distribution is the MultivariateNormalModel$[d, \beta, \sigma^2(X^TX)^{-1}]$. The maximum likelihood estimate of the linear combination $c^T\beta$ of the vector of regression coefficients β is $c^T\beta = c^Tb = c^T(X^TX)^{-1}X^Ty$, the distribution of which is NormalModel$[c^T\beta, \sigma\sqrt{c^T(X^TX)^{-1}c}]$. The MLE of the variance has the form

$$\hat{\sigma}^2 = \frac{(y - X\hat{\beta})^T(y - X\hat{\beta})}{n} = \frac{|y - X\hat{\beta}|^2}{n} = \frac{SS_E}{n} \qquad (3.16)$$

and it is statistically independent of $\hat{\beta}$ such that the distribution of $n\hat{\sigma}^2/\sigma^2$ is χ^2-distribution with $n - d$ degrees of freedom.

To prove these statements we note that the log-likelihood function has the form

$$\ell(\beta, \sigma; y) = -\frac{n}{2}\ln(2\pi) - n\ln(\sigma) - \frac{1}{2\sigma^2}(y - X\beta)^T(y - X\beta). \qquad (3.17)$$

For a fixed β the derivative with respect to σ is

$$\frac{\partial\ell(\beta, \sigma; y)}{\partial\sigma} = -\frac{n}{\sigma} + \frac{1}{\sigma^3}(y - X\beta)^T(y - X\beta)$$

which is a decreasing function of σ obtaining the value zero at

$$\tilde{\sigma}_\beta = \sqrt{\frac{(y - X\beta)^T(y - X\beta)}{n}}.$$

Inserting this into Equation (3.17) one gets

$$\ell(\beta, \tilde{\sigma}_\beta; y) = -\frac{n}{2}\ln(2\pi) - \frac{n}{2}\ln\left(\frac{(y - X\beta)^T(y - X\beta)}{n}\right) - \frac{n}{2},$$

which is maximized at that value of β that minimizes the sum of squared errors.

From the multivariate normal distribution of y and the fact that the MLE $\hat{\beta}$ and the vector of residuals $e = y - X\hat{\beta} = (I - X(X^TX)^{-1}X^T)y$ are linear functions of y follows that they have joint (degenerate) multivariate distribution. Because the covariance matrix between $\hat{\beta}$ and $y - X\hat{\beta}$ is

$$\begin{aligned}
\mathrm{cov}(\hat{\beta}, y - X\hat{\beta}) &= (X^TX)^{-1}X^T\mathrm{var}(y)(I - X(X^TX)^{-1}X) \\
&= \sigma^2(X^TX)^{-1}X^T(I - X(X^TX)^{-1}X^T) \\
&= 0,
\end{aligned}$$

$\hat{\beta}$ and $y - X\hat{\beta}$ are statistically independent and $\hat{\beta}$ has d-dimensional normal distribution with mean vector

$$\mathrm{E}(\hat{\beta}) = (X^TX)^{-1}X^T\mathrm{E}(y) = (X^TX)^{-1}X^TX\beta = \beta \qquad (3.18)$$

and variance matrix

$$\text{var}(\hat{\beta}) = (X^TX)^{-1}X^T\text{var}(y)X(X^TX)^{-1}$$
$$= \sigma^2(X^TX)^{-1}X^TX(X^TX)^{-1}$$
$$= \sigma^2(X^TX)^{-1}. \tag{3.19}$$

The statistic $n\hat{\sigma}^2/\sigma^2$ as a function of $y - X\hat{\beta}$ is statistically independent of $\hat{\beta}$ and has the form

$$n\hat{\sigma}^2/\sigma^2 = \frac{(y - X\hat{\beta})^T(y - X\hat{\beta})}{\sigma^2}$$
$$= \frac{1}{\sigma^2}y^T(I - X(X^TX)^{-1}X^T)y$$
$$= \left(\frac{\varepsilon}{\sigma}\right)^T (I - X(X^TX)^{-1}X^T)\left(\frac{\varepsilon}{\sigma}\right).$$

Now the rank of the matrix $I - X(X^TX)^{-1}X^T$ is

$$\text{rank}(I - X(X^TX)^{-1}X^T) = n - \text{rank}(X(X^TX)^{-1}X^T)$$
$$= n - \text{rank}((X^TX)^{-1}X^TX)$$
$$= n - \text{rank}(I_d)$$
$$= n - d$$

and thus the distribution of $n\hat{\sigma}^2/\sigma^2$ is χ^2-distribution with $n - d$ degrees of freedom because ε/σ is a vector of statistically independent standard normal components.

Example 3.5 Tensile strength data (cont'd)

The log-likelihood function has the form

$$(-10.\beta_1^2 - 46044.\beta_2^2 + 313.6\beta_1 - 1346.\beta_1\beta_2 + 21433.\beta_2 - 2499.17)/\sigma^2 - 20\ln(\sigma)$$

and the MLEs of the vector of regression coefficients and standard deviation are (0.998, 0.218) and 0.693, respectively. ▯

Example 3.6 Fuel consumption data (cont'd)

The log-likelihood function has the form

$$(- 5.\beta_1^2 - 259.525\beta_2^2 - 1948.82\beta_3^2 + 177.35\beta_1 + 56.3\beta_1\beta_2 - 1129.32\beta_2$$
$$- 174.3\beta_1\beta_3 + 1010.78\beta_2\beta_3 + 3221.31\beta_3 - 1627.48)/\sigma^2 - 10\ln(\sigma)$$

and the MLEs of the vector of regression coefficients and standard deviation are (11.934, -0.629, 0.123) and 1.026, respectively. ▯

3.3   Test of linear hypotheses

In a GLM, the uniformly best α-level critical region for the linear hypothesis $H : c^T\beta = a$ with respect to the set of alternative hypotheses $H_A : c^T\beta > a(< a)$ is

$$t_H(y) = \frac{c^T\hat{\beta} - a}{s\sqrt{c^T(X^TX)^{-1}c}} > t_{1-\alpha}[n-d] \ (< t_\alpha[n-d]), \qquad (3.20)$$

where $s^2 = n\hat{\sigma}^2/(n-d)$ is the minimum variance unbiased estimate of the scale parameter (variance) and $t_\alpha[\nu]$ is the αth quantile of the Student t-distribution with ν degrees of freedom. This test is also the likelihood ratio test. The likelihood ratio test of the hypothesis $H : c^T\beta = a$ with respect to the set of alternative hypotheses $H_A : c^T\beta \neq a$ is

$$|t_H(y)| = \frac{|c^T\hat{\beta} - a|}{s\sqrt{c^T(X^TX)^{-1}c}} > t_{1-\alpha/2}[n-d]. \qquad (3.21)$$

If A is a given $q \times d$ matrix with linearly independent rows, then the likelihood ratio test statistic of the hypothesis $H : A\beta = a$ with respect to the set of alternative hypotheses $H_A : A\beta \neq a$ has the form

$$F = \frac{(A\hat{\beta} - a)^T(A(X^TX)^{-1}A^T)^{-1}(A\hat{\beta} - a)/q}{s^2} \qquad (3.22)$$

and its distribution is F-distribution with q numerator and $n-d$ denominator degrees of freedom.

To show that F gives the likelihood ratio test one must find the maximum point of the log-likelihood function under the constraint $H : A\beta = a$. The log-likelihood function of the model has the form

$$\ell(\beta, \sigma; y) = -\frac{n}{2}\ln(2\pi) - n\ln(\sigma) - \frac{(y - X\beta)^T(y - X\beta)}{2\sigma^2}.$$

The Lagrangean function of the constrained optimization function is

$$-\frac{n}{2}\ln(2\pi) - n\ln(\sigma) - \frac{(y - X\beta)^T(y - X\beta)}{2\sigma^2} + \lambda^T(A\beta - a),$$

which gives the following equations for the optimum point:

$$\frac{(y - X\beta)^TX}{\sigma^2} + \lambda^TA = 0,$$

$$-\frac{n}{\sigma} + \frac{(y - X\beta)^T(y - X\beta)}{\sigma^3} = 0,$$

$$A\beta - a = 0.$$

From the first equation by solving for β one gets

$$\beta = (X^{\mathsf{T}}X)^{-1}(X^{\mathsf{T}}y + \sigma^2 A^{\mathsf{T}}\lambda) = \hat{\beta} + \sigma^2(X^{\mathsf{T}}X)^{-1}A^{\mathsf{T}}\lambda,$$

and then inserting this into the third equation and solving for λ one gets

$$\lambda = -\frac{1}{\sigma^2}(A(X^{\mathsf{T}}X)^{-1}A^{\mathsf{T}})^{-1}(A\hat{\beta} - a)$$

and so the constrained MLE of β is

$$\hat{\beta}_{\mathrm{H}} = \hat{\beta} - (X^{\mathsf{T}}X)^{-1}A^{\mathsf{T}}(A(X^{\mathsf{T}}X)^{-1}A^{\mathsf{T}})^{-1}(A\hat{\beta} - a).$$

From the second equation one gets by solving for σ the constrained estimate of the variance

$$\begin{aligned}
\hat{\sigma}_{\mathrm{H}}^2 &= \frac{(y - X\hat{\beta}_{\mathrm{H}})^{\mathsf{T}}(y - X\hat{\beta}_{\mathrm{H}})}{n} \\
&= \frac{(y - X\hat{\beta})^{\mathsf{T}}(y - X\hat{\beta})}{n} + \frac{(A\hat{\beta} - a)^{\mathsf{T}}(A(X^{\mathsf{T}}X)^{-1}A^{\mathsf{T}})^{-1}(A\hat{\beta} - a)}{n} \\
&= \frac{SS_{\mathrm{E}}}{n} + \frac{(A\hat{\beta} - a)^{\mathsf{T}}(A(X^{\mathsf{T}}X)^{-1}A^{\mathsf{T}})^{-1}(A\hat{\beta} - a)}{n}.
\end{aligned}$$

Thus, the likelihood function of the hypothesis H : $A\beta = d$ is

$$\ell_{\mathrm{H}}(d; y) = -\frac{n}{2}\ln\left(\frac{SS_{\mathrm{E}}}{n} + \frac{(A\hat{\beta} - a)^{\mathsf{T}}(A(X^{\mathsf{T}}X)^{-1}A^{\mathsf{T}})^{-1}(A\hat{\beta} - a)}{n}\right) - \frac{n}{2}.$$

Two important special cases are $A = I$ and $A = c^{\mathsf{T}}$, where c is a given d-dimensional vector. The likelihood functions of these special cases are

$$\ell_I(d; y) = -\frac{n}{2}\ln\left(\frac{SS_{\mathrm{E}}}{n} + \frac{(\hat{\beta} - a)^{\mathsf{T}}X^{\mathsf{T}}X(\hat{\beta} - a)}{n}\right) - \frac{n}{2} \qquad (3.23)$$

and

$$\ell_c(a; y) = -\frac{n}{2}\ln\left(\frac{SS_{\mathrm{E}}}{n} + \frac{(c^{\mathsf{T}}\hat{\beta} - a)^2}{n(c^{\mathsf{T}}(X^{\mathsf{T}}X)^{-1}c)}\right) - \frac{n}{2}. \qquad (3.24)$$

Similar considerations give as the global maximum value of the log-likelihood function

$$\ell(\hat{\beta}, \hat{\sigma}; y) = -\frac{n}{2}\ln\left(\frac{SS_{\mathrm{E}}}{n}\right) - \frac{n}{2}.$$

The likelihood ratio statistic of the hypothesis H : $A\beta = a$ has the following form

$$\begin{aligned}
w_{\mathrm{H}}(y) &= 2(\ell(\hat{\beta}, \hat{\sigma}; y) - \ell(\hat{\beta}_{\mathrm{H}}, \hat{\sigma}_{\mathrm{H}}; y)) \\
&= n\ln\left(1 + \frac{(A\hat{\beta} - a)^{\mathsf{T}}(A(X^{\mathsf{T}}X)^{-1}A^{\mathsf{T}})^{-1}(A\hat{\beta} - a)}{(y - X\hat{\beta})^{\mathsf{T}}(y - X\hat{\beta})}\right) \\
&= n\ln\left(1 + \frac{q}{n - d}F_{\mathrm{H}}(y)\right), \qquad (3.25)
\end{aligned}$$

where $F_{\mathrm{H}}(y)$ is F-statistic

$$F_{\mathrm{H}}(y) = \frac{(A\hat{\beta} - a)^{\mathrm{T}}(A(X^{\mathrm{T}}X)^{-1}A^{\mathrm{T}})^{-1}(A\hat{\beta} - a)/q}{(y - X\hat{\beta})^{\mathrm{T}}(y - X\hat{\beta})/(n - d)}$$

$$= \frac{\frac{SS_{\mathrm{H}}}{q}}{\frac{SS_E}{n-d}} = \frac{SS_{\mathrm{H}}}{qs^2}. \tag{3.26}$$

In this expression SS_{H} and SS_{E} denote, respectively, the hypothesis sum of squares and the residual sum of squares

$$SS_{\mathrm{H}} = (A\hat{\beta} - a)^{\mathrm{T}}(A(X^{\mathrm{T}}X)^{-1}A^{\mathrm{T}})^{-1}(A\hat{\beta} - a). \tag{3.27}$$

and

$$SS_{\mathrm{E}} = (y - X\hat{\beta})^{\mathrm{T}}(y - X\hat{\beta}) = (n - d)s^2. \tag{3.28}$$

Now the sums of squares in the expression of F_{H} are statistically independent because the hypothesis sum of squares is a function of the MLE and the residual sum of squares is a function of residuals. Furthermore, we know from a previous section that SS_E/σ^2 has χ^2-distribution with $n - d$ degrees of freedom. In that section it was also shown that the MLE has a d-variate normal distribution with mean vector β and variance matrix $\sigma^2(X^{\mathrm{T}}X)^{-1}$. Thus, under the hypothesis H : $A\beta = a$ the statistic $A\hat{\beta}$ has q-variate normal distribution with mean vector a and variance matrix $\sigma^2 A(X^{\mathrm{T}}X)^{-1}A^{\mathrm{T}}$. The hypothesis sum of squares can be written in the form

$$SS_{\mathrm{H}} = (A\hat{\beta} - a)^{\mathrm{T}}(A(X^{\mathrm{T}}X)^{-1}A^{\mathrm{T}})^{-1}(A\hat{\beta} - a)$$
$$= \sigma^2(A\hat{\beta} - \mathrm{E}(A\hat{\beta}))(\mathrm{var}(A\hat{\beta}))^{-1}(A\hat{\beta} - \mathrm{E}(A\hat{\beta})),$$

from which follows that

$$SS_{\mathrm{H}}/\sigma^2 \sim \chi^2[q]\text{distribution}.$$

Finally

$$F_{\mathrm{H}}(y) = \frac{SS_{\mathrm{H}}/q}{SS_E/(n - d)} = \frac{\left(\frac{SS_{\mathrm{H}}}{\sigma^2}\right)/q}{\left(\frac{SS_E}{\sigma^2}\right)/(n - d)}, \tag{3.29}$$

where the numerator and denominator are statistically independent such that distributions of $\frac{SS_{\mathrm{H}}}{\sigma^2}$ and $\frac{SS_E}{\sigma^2}$ are χ^2-distribution with q degrees of freedom and χ^2-distribution with $n - d$ degrees of freedom, respectively. From this follows that the distribution of F_{H} is F-distribution with q numerator and $n - d$ denominator degrees of freedom.

Because the likelihood ratio statistic of the hypothesis H : $A\beta = a$ is a monotone function of the F-statistic, the observed significance level for the hypothesis H : $A\beta = a$ calculated from the observation y_{obs} and the model is equal to

$$\Pr(w_H(y) \geq w_H(y_{\text{obs}}); \beta, \sigma) = \Pr(F_H(y) \geq F_H(y_{\text{obs}}); \beta, \sigma)$$
$$= 1 - F_{F[q,n-d]\text{distribution}}(F_H(y_{\text{obs}})),$$

as in the linear model, when the hypothesis H : $A\beta = a$ is satisfied, the distribution of the F-statistic $F_H(y)$ is F-distribution with q numerator and $n - d$ denominator degrees of freedom.

In the special case $A = c^T$ the F-statistic takes the form

$$F_H(y) = \frac{(c^T\hat{\beta} - a)^2}{s^2 c^T (X^T X)^{-1} c} = t_H(y)^2, \tag{3.30}$$

where $t_H(y)$ is the Student t-statistic.

Example 3.7 Fuel consumption data (cont'd)

Consider the hypothesis that neither of the regressors has an effect on the amount of fuel consumption. This hypothesis can be expressed in the form

$$\begin{pmatrix} 0\ 1\ 0 \\ 0\ 0\ 1 \end{pmatrix} \beta = \begin{pmatrix} 0 \\ 0 \end{pmatrix}$$

that is, $\beta_2 = 0$ and $\beta_3 = 0$. The observed significance level calculated from the F-ratio distribution with 2 and 7 degrees of freedom is 0.000274.   ☐

3.4   Confidence regions and intervals

3.4.1   Joint confidence regions for finite sets of linear combinations

Although in testing the acceptability of a hypothesis in general, no reference to the theory of estimation is required, at least in this important class of cases, beyond the stipulation that expectations can be fitted efficiently to the data, yet when the general hypothesis is found to be acceptable, and accepting it as true, we proceed to the next step of discussing the bearing of the observational record upon the problem of discriminating among the various possible values of the parameter, we are discussing the theory of estimation itself. (Fisher 1956, p. 49)

This quotation from Fisher's book *Statistical Methods and Scientific Inference* (1956) and the one in the beginning of Section 1.7 make it clear that for Fisher "estimation" meant finding out what values of parameter are supported by statistical evidence and what are not supported by it, that is, region or interval estimation.

In a general linear model the $(1 - \alpha)$-level confidence region for a finite set of linear combinations $\psi = A\beta$ of regression coefficients has the form

$$\{\psi : \ (\psi - A\hat{\beta})^{\mathrm{T}}(A(X^{\mathrm{T}}X)^{-1}A^{\mathrm{T}})^{-1}(\psi - A\hat{\beta}) \leq qF_{1-\alpha}[q, n - d]s^2\}, \quad (3.31)$$

where $F_{1-\alpha}[q, n - d]$ is $(1 - \alpha)^{\mathrm{th}}$ quantile of F-distribution with q numerator and $n - d$ denominator degrees of freedom..

To prove this we note that in Section 3.3 the likelihood function of the interest function $\psi = A\beta$ was shown to be equal to

$$\ell_P(\psi; y) =$$
$$-\frac{n}{2}\ln\left(\frac{(y - X\hat{\beta})^{\mathrm{T}}(y - X\hat{\beta})}{n} + \frac{(A\hat{\beta} - \psi)^{\mathrm{T}}(A(X^{\mathrm{T}}X)^{-1}A^{\mathrm{T}})^{-1}(A\hat{\beta} - \psi)}{n}\right) - \frac{n}{2}.$$

Thus, the profile likelihood-based confidence region has the form

$$\{\psi : \ell_P(\psi; y_{\mathrm{obs}}) \geq \ell(\hat{\beta}, \hat{\sigma}; y_{\mathrm{obs}}) - c/2\}$$
$$= \left\{\psi : n\ln\left(1 + \frac{q}{n - d}F_{\mathrm{H}}(y)\right) \leq c\right\}$$
$$= \left\{\psi : F_{\psi}(y) \leq \frac{n - d}{q}\left(e^{\frac{c}{n}} - 1\right)\right\},$$

where
$$F_{\psi}(y) = \frac{(A\hat{\beta} - \psi)^{\mathrm{T}}(A(X^{\mathrm{T}}X)^{-1}A^{\mathrm{T}})^{-1}(A\hat{\beta} - \psi)/q}{s^2}.$$

Because the F-statistic has the F-distribution with q numerator and $n - d$ denominator degrees of freedom as distribution the confidence level of the above confidence region will be equal to $1 - \alpha$, if we choose c such that

$$\frac{n - d}{q}\left(e^{\frac{c}{n}} - 1\right) = F_{1-\alpha}[q, n - d],$$

that is,

$$c = n\ln\left(1 + \frac{q}{n - d}F_{1-\alpha}[q, n - d]\right).$$

The $(1-\alpha)$-level confidence region for the whole regression coefficient vector has the form

$$\{\beta : \ (\beta - \hat{\beta})^{\mathrm{T}}X^{\mathrm{T}}X(\beta - \hat{\beta}) \leq dF_{1-\alpha}[d, n - d]s^2\}.$$

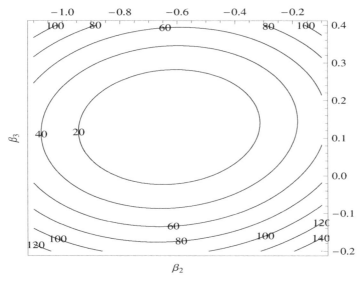

FIGURE 3.4:   Fuel consumption data: Confidence region.

Example 3.8 Fuel consumption data (cont'd)

Consider the coefficients of two regressors, that is, $\psi_1 = \beta_2$ and $\psi_2 = \beta_3$. Then the 0.95-level confidence region is

$$\{(\beta_2, \beta_3) : 202.081\beta_2^2 - 58.942\beta_2\beta_3 + 261.679\beta_2 + 859.581\beta_3^2$$
$$- 260.203\beta_3 + 99.1244 \le 9.97967\}.$$

Figure 3.4 gives a contour plot of it.   ☐

3.4.2   Separate confidence intervals for linear combinations

Most practical problems of parametric inference, it seems to me, fall in the category of "inference in the presence of nuisance parameters," with the acknowledgement that various functions $g(\theta)$, which might be components of a parameter vector, are simultaneous interest. That is, in practice, we may wish to make inferences about each of many such functions $g(\theta)$ in turn, keeping in mind that joint inferences are being made, yet requiring separate statements for ease of contemplation. (Kass in Reid 1988, p. 236)

The $(1 - \alpha)$-level confidence interval for the linear combination $\psi = c^{\mathrm{T}}\beta$

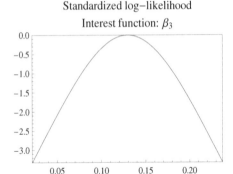

FIGURE 3.5:   Fuel consumption data: Plot of the log-likelihood of β_3.

has the form

$$\left\{\psi : \frac{(\psi - c^{\mathrm{T}}\hat{\beta})^2}{c^{\mathrm{T}}(X^{\mathrm{T}}X)^{-1}c} \leq F_{1-\alpha}[1, n-d]s^2\right\} = c^{\mathrm{T}}\hat{\beta}\mp t_{1-\alpha/2}[n-d]s\sqrt{c^{\mathrm{T}}(X^{\mathrm{T}}X)^{-1}c}.$$

(3.32)

The result follows immediately from the results of the previous section by taking $\psi = c^{\mathrm{T}}\beta$.

Example 3.9 Fuel consumption data (cont'd)

Consider the regression coefficient β_3, which can be written in the form $\beta_3 = \psi = (\,0\,0\,1\,)\,\beta$. An exact 95%-level confidence interval based on the appropriate Student t-distribution with 7 degrees of freedom is $\{0.0306, 0.229\}$. Figure 3.5 shows the relative log-likelihood function of β_3.   ⬜

Example 3.10 Red clover plants data

The following example studies the effect of bacteria on the nitrogen content of red clover plants. The treatment factor is bacteria strain, and it has six levels. Five of the six levels consist of five different *Rhizobium trifolii* bacteria cultures combined with a composite of five *Rhizobium meliloti* strains. The sixth level is a composite of the five *Rhizobium trifolii* strains with the composite of the *Rhizobium meliloti*. Red clover plants are inoculated with the treatments, and nitrogen content is later measured in milligrams. The data are derived from an experiment by Erdman (1946) and are analyzed in Chapters 7 and 8 of Steel and Torrie (1980). The columns in Table A.18 on page 265 in Appendix A contain values of the treatment factor and nitrogen content. The effect of treatment on nitrogen content is under investigation. Figure 3.6 plots the nitrogen content values for all treatments. The treatment seems to affect the nitrogen content.

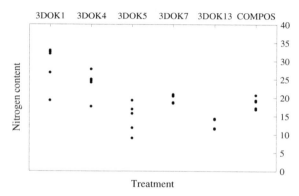

FIGURE 3.6: Red clover plants data: Nitrogen contents for treatment groups.

Taking cell means as regression coefficients results in the following model matrix

$$X = \begin{pmatrix} 1\,0\,0\,0\,0\,0 \\ 1\,0\,0\,0\,0\,0 \\ 1\,0\,0\,0\,0\,0 \\ 1\,0\,0\,0\,0\,0 \\ 1\,0\,0\,0\,0\,0 \\ 0\,1\,0\,0\,0\,0 \\ 0\,1\,0\,0\,0\,0 \\ \vdots\,\vdots\,\vdots\,\vdots\,\vdots\,\vdots \\ 0\,0\,0\,0\,1\,0 \\ 0\,0\,0\,0\,1\,0 \\ 0\,0\,0\,0\,0\,1 \\ 0\,0\,0\,0\,0\,1 \\ 0\,0\,0\,0\,0\,1 \\ 0\,0\,0\,0\,0\,1 \\ 0\,0\,0\,0\,0\,1 \end{pmatrix}.$$

Table 3.2 contains estimates and exact 0.95-level confidence intervals, that is, t-intervals for all pairwise comparisons. ▯

Example 3.11 Waste data

The data are from an experiment designed to study the effect of temperature (at low, medium, and high levels) and environment (five different environments) on waste output in a manufacturing plant (Westfall *et al.* 1999). The columns in Table A.25 on page 270 in Appendix A contain values of temperature, environment, and waste output. The dependence of waste output on temperature and environment is under investigation. Figure 3.7 gives a

TABLE 3.2:   Red clover plants data: Estimates
and exact 0.95-level confidence intervals (t-intervals)
for pairwise differences of means.

Difference	Estimate	Lower limit	Upper limit
$\mu_1 - \mu_2$	4.84	0.358	9.322
$\mu_1 - \mu_3$	14.18	9.698	18.662
$\mu_1 - \mu_4$	8.90	4.418	13.382
$\mu_1 - \mu_5$	15.56	11.078	20.042
$\mu_1 - \mu_6$	10.12	5.638	14.602
$\mu_2 - \mu_3$	9.34	4.858	13.822
$\mu_2 - \mu_4$	4.06	-0.423	8.542
$\mu_2 - \mu_5$	10.72	6.238	15.202
$\mu_2 - \mu_6$	5.28	0.798	9.762
$\mu_3 - \mu_4$	-5.28	-9.762	-0.798
$\mu_3 - \mu_5$	1.38	-3.102	5.862
$\mu_3 - \mu_6$	-4.06	-8.542	0.422
$\mu_4 - \mu_5$	6.66	2.178	11.142
$\mu_4 - \mu_6$	1.22	-3.262	5.702
$\mu_5 - \mu_6$	-5.44	-9.922	-0.958

plot of the values of waste output for different temperatures and environments
showing that both factors might affect the waste output.

Taking cell means as regression coefficients results in the following model
matrix

$$X = \begin{pmatrix} 1\,0\,0\,0\,0\,0\,0\,0\,0\,0\,0\,0\,0\,0 \\ 1\,0\,0\,0\,0\,0\,0\,0\,0\,0\,0\,0\,0\,0 \\ 0\,1\,0\,0\,0\,0\,0\,0\,0\,0\,0\,0\,0\,0 \\ 0\,1\,0\,0\,0\,0\,0\,0\,0\,0\,0\,0\,0\,0 \\ \vdots\,\vdots\,\vdots\,\vdots\,\vdots\,\vdots\,\vdots\,\vdots\,\vdots\,\vdots\,\vdots\,\vdots\,\vdots\,\vdots \\ 0\,0\,0\,0\,0\,0\,0\,0\,0\,0\,0\,0\,1\,0 \\ 0\,0\,0\,0\,0\,0\,0\,0\,0\,0\,0\,0\,1\,0 \\ 0\,0\,0\,0\,0\,0\,0\,0\,0\,0\,0\,0\,0\,1 \\ 0\,0\,0\,0\,0\,0\,0\,0\,0\,0\,0\,0\,0\,1 \end{pmatrix}.$$

Table 3.3 contains estimates and exact 0.95-level confidence intervals, that is,
t-intervals for a set of interaction contrasts spanning the interaction space. □

3.5   Model checking

Applying the theory of Section 2.6 to linear regression models first gives

$$\mu_i(\beta) = \eta_i(\beta) = x_i^T \beta$$

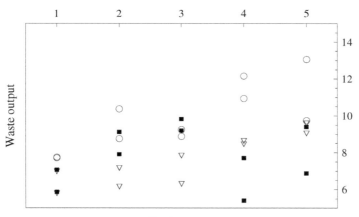

FIGURE 3.7: Waste data: Waste ouput for environment groups. Symbols ■, ▽, and ○ denote low, medium, and high temperature, respectively.

with

$$\ell(\mu_i, \sigma; y_i) = -\frac{1}{2}\ln(2\pi) - \ln(\sigma) - \frac{1}{2\sigma^2}(y_i - \mu_i)^2.$$

From this follows that $\tilde{\mu}_i = \tilde{\eta}_i = y_i$ and so

$$d_i = \text{sign}(y_i - \hat{\mu}_i) \left[2\{\ell_i(y_i; y_i, \sigma) - \ell_i(y_i; \hat{\mu}_i, \sigma)\}\right]^{1/2}$$
$$= \frac{y_i - \hat{\mu}_i}{\sigma}, \tag{3.33}$$

where

$$\hat{\mu}_i = \mu_i(\hat{\beta}) = x_i^\mathrm{T}\hat{\beta}.$$

The hat matrix is equal to

$$H(\hat{\beta}) = W(\hat{\beta})^{1/2}X(\hat{\beta})(X(\hat{\beta})^\mathrm{T}W(\hat{\beta})X(\hat{\beta}))^{-1}X(\hat{\beta})^\mathrm{T}W(\hat{\beta})^{1/2}$$
$$= X(X^\mathrm{T}X)^{-1}X^\mathrm{T},$$

TABLE 3.3: Waste data: Estimates and exact 0.95-level confidence intervals (t-intervals) of interaction contrasts.

Interaction contrast	Estimate	Lower limit	Upper limit
$\mu_{11} - \mu_{12} - \mu_{21} + \mu_{22}$	-1.78	-4.926	1.366
$\mu_{11} - \mu_{13} - \mu_{21} + \mu_{23}$	-2.36	-5.511	0.781
$\mu_{11} - \mu_{14} - \mu_{21} + \mu_{24}$	2.08	-1.066	5.226
$\mu_{11} - \mu_{15} - \mu_{21} + \mu_{25}$	1.26	-1.886	4.406
$\mu_{11} - \mu_{12} - \mu_{31} + \mu_{32}$	-0.22	-3.366	2.926
$\mu_{11} - \mu_{13} - \mu_{31} + \mu_{33}$	-1.71	-4.861	1.431
$\mu_{11} - \mu_{14} - \mu_{31} + \mu_{34}$	3.72	0.574	6.866
$\mu_{11} - \mu_{15} - \mu_{31} + \mu_{35}$	1.99	-1.156	5.136

because

$$X(\beta) = \frac{\partial \eta(\beta)}{\partial \beta^{\mathrm{T}}} = \frac{\partial (X\beta)}{\partial \beta^{\mathrm{T}}} = \left(\frac{\partial x_i^{\mathrm{T}} \beta}{\partial \beta_j} \right)_{n \times d} = (x_{ij})_{n \times d} = X$$

and

$$W(\beta, \sigma) = \mathrm{diag} \left(\mathrm{E} \left(-\frac{\partial^2 \ell_i (\beta, \sigma; y_i)}{\partial \mu_i^2} \right) \right)$$

$$= \mathrm{diag} \left(\left(\frac{1}{\sigma^2}, \ldots, \frac{1}{\sigma^2} \right)_n \right)$$

$$= \frac{1}{\sigma^2} I_n.$$

The standardized deviance residual thus has the form

$$r_{Di} = \frac{d_i}{\sqrt{1 - h_{ii}}} = \frac{y_i - \hat{\mu}_i}{\sigma \sqrt{1 - h_{ii}}}. \tag{3.34}$$

Now

$$u_i(\beta, \sigma) = \frac{\partial \ell_i(\beta, \sigma)}{\partial \mu_i} = \frac{y_i - \mu_i}{\sigma^2}$$

and so

$$r_{Pi} = \frac{u_i(\hat{\beta})}{\sqrt{w_i(\hat{\beta})(1 - h_{ii})}} = \frac{y_i - \hat{\mu}_i}{\sigma^2 \sqrt{\frac{1}{\sigma^2}(1 - h_{ii})}}$$

$$= \frac{y_i - \hat{\mu}_i}{\sigma \sqrt{(1 - h_{ii})}} = r_{Di}.$$

From this equality follows that

$$r_i^* = r_{Di} + \frac{1}{r_{Di}} \ln \left(\frac{r_{Pi}}{r_{Di}} \right) = r_{Di} = r_{Pi} \ \forall i = 1, \ldots, n. \tag{3.35}$$

Because σ is unknown, the actual residuals will be obtained by replacing σ by its estimate, that is, by the MLE $\hat{\sigma}$ and so one gets

$$r_i^* = r_{Di} = r_{Pi} = \frac{y_i - \hat{\mu}_i}{\hat{\sigma} \sqrt{(1 - h_{ii})}} \ \forall i = 1, \ldots, n.$$

Asymptotically these residuals have marginal standard normal distribution, but in this case one can say more. Namely using unbiased estimate s instead of $\hat{\sigma}$, that is, considering

$$r_i^* = r_{Di} = r_{Pi} = \frac{y_i - \hat{\mu}_i}{s \sqrt{(1 - h_{ii})}} \ \forall i = 1, \ldots, n \tag{3.36}$$

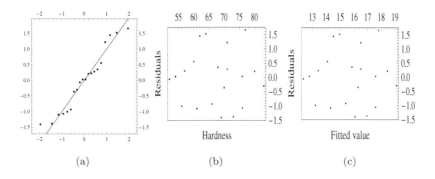

FIGURE 3.8:   Tensile strength data: (a) QQ-plot and plots of residuals against (b) hardness and (c) fitted values.

one gets *studentized residuals* the marginal distribution of which is Student t-distribution with $n - d$ degrees of freedom.

Example 3.12 Tensile strength data (cont'd)

Figure 3.8 gives a quantile-quantile plot along with plots of standardized residuals against hardness and fitted values in the fitted simple linear regression model.   ▯

Example 3.13 Fuel consumption data (cont'd)

Figure 3.9 gives a quantile-quantile plot along with plots of standardized residuals against air temperature, wind velocity, and fitted values in the fitted multiple linear regression model.

3.6   Bibliographic notes

Basic books on linear models are Searle (1971), Seber (1977), and Sen and Srivastava (1990). Davison (2003) contains an excellent account of linear models among other subjects. Cook and Weisberg (1971) is a book on residuals.

Interesting historical material can be found in Laplace (1774), Legendre (1805), Gauss (1809), and "Student" (1908).

In addition, Stigler's book (1986) includes a thorough account of the development of the concept of the linear model.   ⬚

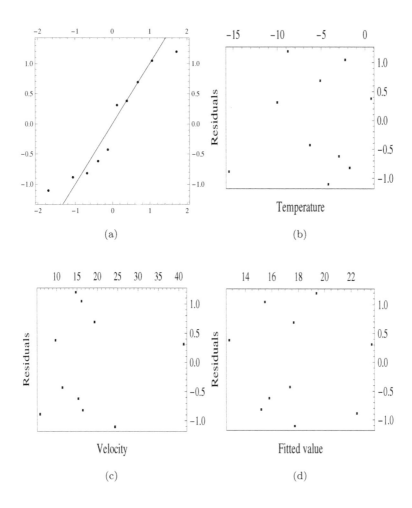

FIGURE 3.9: Fuel consumption data: (a) QQ-plot and plots of residuals against (b) air temperature, (c) wind velocity, and (d) fitted values.

Chapter 4

Nonlinear Regression Model

In this chapter in Section 4.1 the definition of nonlinear regression model is presented. In Section 4.2 estimation of regression parameters is discussed by first considering least squares and then the maximum likelihood method. Then in Section 4.3 the approximate distribution of likelihood ratio statistic is presented and applied in Section 4.4 and Section 4.5 to derive profile likelihood-based regions to whole or part of the vector of the regression parameters and profile likelihood-based confidence interval to any smooth real-valued function of the regression parameters, respectively. In Section 4.6 the likelihood ratio test of a nonlinear hypothesis on the regression parameters is considered. Finally, Section 4.7 contains a discussion of model checking.

4.1 Definition of nonlinear regression model

The nonlinear regression model was defined in Section 2.3 as a univariate and one-parameter GRM with statistically independent normally distributed components of response vector such that means of the components depend nonlinearly on the regression parameters. Classically this is expressed with the formula

$$y_i = f(x_i; \beta) + \varepsilon_i \ \forall i = 1, \ldots, n, \tag{4.1}$$

where the x_is contain values of explaining variables, f is a known function of explaining variables and vector β of the unknown regression parameters, and the ε_is are statistically independent and normally distributed with zero means and common unknown standard deviation σ. The unknown regression parameter vector is assumed to belong to a given open subset B of d-dimensional Euclidean space, that is, $\beta = (\beta_1, \ldots, \beta_d) \in B \subset \mathbb{R}^d$.

From assumptions follows that the likelihood function of the model is

$$L(\beta, \sigma; y) = \prod_{i=1}^{n} \frac{1}{\sqrt{2\pi}\sigma} e^{-\frac{(y_i - f(x_i; \beta))^2}{2\sigma^2}}$$

TABLE 4.1:   Puromycin data.

Treatment	Concentration	Velocity
Treated	0.02	76
Treated	0.02	47
Untreated	0.02	67
Untreated	0.02	51
Treated	0.06	97
Treated	0.06	107
Untreated	0.06	84
Untreated	0.06	86
Treated	0.11	123
Treated	0.11	139
Untreated	0.11	98
Untreated	0.11	115
Treated	0.22	159
Treated	0.22	152
Untreated	0.22	131
Untreated	0.22	124
Treated	0.56	191
Treated	0.56	201
Untreated	0.56	144
Untreated	0.56	158
Treated	1.10	207
Treated	1.10	200
Untreated	1.10	160

and the log-likelihood function, respectively,

$$\ell(\beta, \sigma; y) = -\frac{n}{2} \ln(2\pi) - n \ln(\sigma) - \frac{1}{2\sigma^2} \sum_{i=1}^{n} (y_i - f(x_i; \beta))^2. \qquad (4.2)$$

Example 4.1 Puromycin data

The number of counts per minute of radioactive product from an enzymatic reaction was measured as a function of substrate concentration in parts per million and from these counts the initial rate, or "velocity" of the reaction was calculated. The experiment was conducted once with the enzyme treated with Puromycin and once with the enzyme untreated (Bates and Watts 1988, p. 269). The columns in Table 4.1 contain values of treatment, concentration (ppm), and velocity (counts/ $\min^2$). Figure 4.1 contains a plot of velocity versus concentration for the groups.

Let us consider the *Treated* group. Values of velocity of the enzymatic

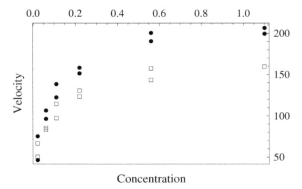

FIGURE 4.1:   Puromycin data: Velocity against concentration for treated (•) and untreated ($\square$).

reaction form response and model matrix is

$$\mathcal{N}X = \begin{pmatrix} 0.02 \\ 0.02 \\ 0.06 \\ \vdots \\ 0.56 \\ 1.10 \\ 1.10 \end{pmatrix}.$$

Assume that velocity depends on the substrate concentration according to the *Michaelis-Menten* equation, that is, the regression function has the form

$$f(x; \beta_1, \beta_2) = \frac{\beta_1 x}{x + \beta_2}$$

and the nonlinear regression model is

$$\mathcal{N}\mathcal{M} = \text{NonlinearRegressionModel}\left[\mathcal{N}X, \frac{\beta_1 x}{x + \beta_2}, \{x\}, \{\beta_1, \beta_2\}\right].$$

$\square$

4.2   Estimate of regression parameters

4.2.1   Least squares estimate (LSE) of regression parameters

Results considered in this section are more general than the assumptions made in Section 4.1 imply. Namely, instead of assuming components of the response vector being statistically independent, it suffices to assume that their mean vector is $(f(x_1; \beta), \ldots, f(x_n; \beta))^{\mathrm{T}}$, they have a common variance σ^2, and they are uncorrelated without any assumption being made about their distributional form. If, however, the normal distribution assumption is made the assumption of uncorrelatedness implies statistical independence. One can even be slightly more general and because this more general case is needed later on, that assumption, namely the assumption of the covariance matrix of the response vector being known up to an unknown scale parameter, is made in this section. Thus, the covariance matrix of the response vector is assumed to have the form

$$\Sigma = \sigma^2 V, \tag{4.3}$$

where V is a known nonsingular positive definite $n \times n$ matrix. The *least squares estimate* (LSE) of the regression parameter vector β is that value of β which minimizes the expression

$$\sum_{i=1}^{n} e_i^2 = \sum_{i=1}^{n} (y_i - f(x_i; \beta))^2$$
$$= (y - f(X; \beta))^{\mathrm{T}} (y - f(X; \beta))$$
$$= |y - f(X; \beta)|^2. \tag{4.4}$$

or in the more general case the expression

$$e^{\mathrm{T}} V^{-1} e = (y - f(X; \beta))^{\mathrm{T}} V^{-1} (y - f(X; \beta)) = S(\beta), \tag{4.5}$$

where x_i is the ith row of the model matrix X and

$$f(X; \beta) = (f(x_1; \beta), \ldots, f(x_n; \beta))^{\mathrm{T}}. \tag{4.6}$$

The LSE b will be obtained as a solution of the following Gauss *normal equations*

$$\tilde{X}(\beta)^{\mathrm{T}} f(X; \beta) = \tilde{X}(\beta)^{\mathrm{T}} y \tag{4.7}$$

or more generally

$$\tilde{X}(\beta)^{\mathrm{T}} V^{-1} f(X; \beta) = \tilde{X}(\beta)^{\mathrm{T}} V^{-1} y, \tag{4.8}$$

where

$$\tilde{X}(\beta) = \frac{\partial f(X; \beta)}{\partial \beta^{\mathrm{T}}} = \left(\frac{\partial f(x_i; \beta)}{\partial \beta_j} \right) \tag{4.9}$$

is an $n \times d$ matrix with the ith row containing partial derivatives of $f(x_i; \beta)$ with respect to the components of the regression parameter vector.

Geometric interpretation of these equations is such that the set $\{f(X; \beta) : \beta \in B\}$ defines a d-dimensional *regression surface* in n-dimensional sample space $\mathbb{R}^n$. The LSE of the regression parameter vector is that vector in the set B for which the corresponding vector $f(X; \beta)$ in the regression surface minimizes the distance between y and $f(X;\beta)$, that is, which gives the point in the regression surface closest to the observed response vector y. The Gauss normal equations come from the fact that given that the regression surface is smooth enough and has a tangent "plane" determined by the columns of the matrix $\tilde{X}(\beta)$, the difference $y - f(X;\beta)$ must be orthogonal to this tangent "plane." The normal equations are nonlinear with respect to the components of the parameter vector and normally need to be solved by some iterative method. One such iterative method is the *Gauss-Newton method*, which is based on approximating the regression surface by the linear surface at the current iteration point β^0, that is, the tangent "plane" at β^0. Thus, using

$$f(X;\beta) \approx f(X;\beta^0) + \tilde{X}(\beta^0)(\beta - \beta^0)$$

normal equations have the form

$$\tilde{X}(\beta^0)^{\mathrm{T}}V^{-1}(f(X;\beta^0) + \tilde{X}(\beta^0)(\beta - \beta^0)) = \tilde{X}(\beta^0)^{\mathrm{T}}V^{-1}y$$

or

$$\tilde{X}(\beta^0)^{\mathrm{T}}V^{-1}\tilde{X}(\beta^0)(\beta - \beta^0) = \tilde{X}(\beta^0)^{\mathrm{T}}V^{-1}(y - f(X;\beta^0)).$$

This gives the solution

$$\beta^1 = \beta^0 + (\tilde{X}(\beta^0)^{\mathrm{T}}V^{-1}\tilde{X}(\beta^0))^{-1}\tilde{X}(\beta^0)^{\mathrm{T}}V^{-1}(y - f(X;\beta^0)) \qquad (4.10)$$

for the next iteration point.

Instead of using the Gauss-Newton method it is customary to use some general or specialized minimization methods, like Newton-Raphson or Levenberg-Marquardt methods. These will not be discussed further because in the following always the general iteration method for maximizing the likelihood function is used. The MLE of the regression parameter vector is equivalent to the LSE and at the same time an estimate to the standard deviation is obtained. When using this iterative method to calculate LSE (MLE) care must be exercised in selection of the starting point for the iteration. There does not exist general rules for the selection of a starting point, but good advice can be found in Bates and Watts (1988).

Using the Gauss-Newton iteration scheme starting from the "true" value of the regression parameter vector β and taking only one step we get

$$b \approx \beta + W(\beta)^{-1}\tilde{X}(\beta)^{\mathrm{T}}V^{-1}\varepsilon,$$

where

$$W(\beta) = \tilde{X}(\beta)^{\mathrm{T}}V^{-1}\tilde{X}(\beta).$$

The mean vector of b is approximately

$$\mathrm{E}(b) \approx \beta + W(\beta)^{-1}\tilde{X}(\beta)^{\mathrm{T}}V^{-1}\mathrm{E}(\varepsilon) = \beta \qquad (4.11)$$

and the variance matrix is approximately

$$\mathrm{var}(b) \approx \mathrm{var}(W(\beta)^{-1}\tilde{X}(\beta)^{\mathrm{T}}V^{-1}\varepsilon) = \sigma^2 W(\beta)^{-1}. \qquad (4.12)$$

Finally, the LSE and the vector of *residuals* $e = y - f(X; b)$ are approximately uncorrelated because

$$
\begin{aligned}
\text{cov}(b, e) &\approx \text{cov}(W(\beta)^{-1}\tilde{X}(\beta)^{\mathrm{T}}V^{-1}\varepsilon, (I - \tilde{X}(\beta)W(\beta)^{-1}\tilde{X}(\beta)^{\mathrm{T}}V^{-1})\varepsilon) \\
&= \sigma^2 W(\beta)^{-1}\tilde{X}(\beta)^{\mathrm{T}}V^{-1}V(I - V^{-1}\tilde{X}(\beta)W(\beta)^{-1}\tilde{X}(\beta)^{\mathrm{T}}) \\
&= \sigma^2 (W(\beta)^{-1}\tilde{X}(\beta)^{\mathrm{T}} - W(\beta)^{-1}\tilde{X}(\beta)^{\mathrm{T}}) \\
&= 0.
\end{aligned}
$$

4.2.2 Maximum likelihood estimate (MLE)

In a nonlinear regression model the MLE $\hat{\beta}$ of the vector of regression parameters β is the same as the LSE b. Because of the approximation

$$
\hat{\beta} \approx \beta + W(\beta)^{-1}\tilde{X}(\beta)^{\mathrm{T}}V^{-1}\varepsilon,
$$

the MLE has approximately d-dimensional normal distribution with mean vector β and variance matrix

$$
\text{var}(\hat{\beta}) \approx \sigma^2 W(\beta)^{-1}. \tag{4.13}
$$

The MLE $\hat{\beta}$ and the vector of residuals $y - f(X; \hat{\beta})$ are approximately statistically independent. These results follow from the form of the log-likelihood function

$$
\ell(\beta, \sigma; y) = -\frac{n}{2}\ln(2\pi) - n\ln(\sigma) - \frac{1}{2}\ln(\det(V)) - \frac{1}{2\sigma^2}S(\beta).
$$

First, because for a fixed β the derivative of the log-likelihood function with respect to σ is

$$
\frac{\partial \ell(\beta, \sigma; y)}{\partial \sigma} = -\frac{n}{\sigma} + \frac{1}{\sigma^3}S(\beta)
$$

which is a decreasing function of σ obtaining the value zero at

$$
\tilde{\sigma}_\beta = \sqrt{\frac{S(\beta)}{n}}.
$$

Inserting this into the log-likelihood function one gets

$$
\ell(\beta, \tilde{\sigma}_\beta; y) = -\frac{n}{2}\ln(2\pi) - \frac{n}{2}\ln\left(\frac{S(\beta)}{n}\right) - \frac{1}{2}\ln(\det(V)) - \frac{n}{2}
$$

which is maximized at that value of β which minimizes the sum of squared errors. Finally, the MLE of σ is

$$
\hat{\sigma} = \sqrt{\frac{S(\hat{\beta})}{n}}.
$$

Now because the MLE $\hat{\beta}(= b)$ and the residual vector $y - f(X; \hat{\beta})$ are approximately uncorrelated they are also approximately statistically independent.

The statistic $n\hat{\sigma}^2/\sigma^2$ as a function of $y - X\hat{\beta}$ is approximately statistically independent of $\hat{\beta}$ and has approximately the form

$$n\hat{\sigma}^2/\sigma^2 = \frac{(y - f(X; \hat{\beta}))^{\mathrm{T}} V^{-1} (y - f(X; \hat{\beta}))}{\sigma^2}$$

$$\approx \frac{1}{\sigma^2} \varepsilon^{\mathrm{T}} (V^{-1} - V^{-1} \tilde{X}(\beta) W(\beta)^{-1} \tilde{X}(\beta)^{\mathrm{T}} V^{-1}) \varepsilon$$

$$= \left(\frac{\varepsilon}{\sigma}\right)^{\mathrm{T}} (V^{-1} - V^{-1} \tilde{X}(\beta) W(\beta)^{-1} \tilde{X}(\beta)^{\mathrm{T}} V^{-1}) \frac{\varepsilon}{\sigma}.$$

Now the rank of the matrix

$$V^{-1} - V^{-1} \tilde{X}(\beta) W(\beta)^{-1} \tilde{X}(\beta)^{\mathrm{T}} V^{-1}$$

is that of the matrix

$$W = I - \tilde{X}(\beta) W(\beta)^{-1} \tilde{X}(\beta)^{\mathrm{T}} V^{-1},$$

which is

$$\mathrm{rank}(W) = n - \mathrm{rank}(\tilde{X}(\beta) W(\beta)^{-1} \tilde{X}(\beta)^{\mathrm{T}} V^{-1})$$

$$= n - \mathrm{rank}(W(\beta) W(\beta)^{-1})$$

$$= n - \mathrm{rank}(I_d) = n - d$$

and thus $n\hat{\sigma}^2/\sigma^2$ has approximately χ^2-distribution with $n - d$ degrees of freedom because ε/σ is a vector of the statistically independent standard normal components.

Example 4.2 Puromycin data (cont'd)

The regression function is monotonely increasing and has asymptote β_1. Thus, the maximum value 207 of velocities gives a starting value for β_1. At concentration β_2 the regression function takes value $\beta_1/2$ and so the value 0.06 of concentration for which velocity has the value $\beta_1/2$ gives the starting value for β_2. The linear model

$$\mathcal{LM} = \mathrm{LinearRegressionModel} \left[\begin{pmatrix} 1 \\ \vdots \\ 1 \end{pmatrix} \right]$$

for residuals calculated from "known" values 207 and 0.06 gives a starting value 10.039 for the standard deviation. MLEs of the vector of regression parameters and standard deviation are (212.7, 0.0641) and 9.981, respectively. This shows that starting values were rather near the corresponding values of the MLE. ∎

4.3   Approximate distribution of LRT statistic

Let us consider a submodel of a nonlinear regression model such that the regression parameter vector is constrained to belong to a subset of B defined by the condition $g(\beta) = 0$, where g is a given smooth q-dimensional vector valued function $(1 \le q \le d)$. Denote this subset by Γ. It is a $(d-q)$-dimensional "surface" contained in B. Now the likelihood ratio statistic of the statistical hypothesis $\mathrm{H} : g(\beta) = 0$, that is, $\mathrm{H} : \beta \in \Gamma$ has the form

$$2\{\ell(\hat{\beta}, \hat{\sigma}; y_{\mathrm{obs}}) - \ell(\tilde{\beta}, \tilde{\sigma}; y_{\mathrm{obs}})\} = 2\left\{-\frac{n}{2}\ln\left(\frac{S(\hat{\beta})}{n}\right) - \left(-\frac{n}{2}\ln\left(\frac{S(\tilde{\beta})}{n}\right)\right)\right\}$$

$$= n\ln\left(\frac{S(\tilde{\beta})}{S(\hat{\beta})}\right)$$

$$= n\ln\left(1 + \frac{S(\tilde{\beta}) - S(\hat{\beta})}{S(\hat{\beta})}\right). \qquad (4.14)$$

In the case of a linear model

$$y = X\beta + \varepsilon$$

under the usual assumptions it was shown in Section 3.3 that for a linear hypothesis $\mathrm{H} : A\beta = a$ or $\mathrm{H} : A\beta - a = 0$ the likelihood ratio statistic is

$$2\{\ell(\hat{\beta}, \hat{\sigma}; y_{\mathrm{obs}}) - \ell(\tilde{\beta}, \tilde{\sigma}; y_{\mathrm{obs}})\} = n\ln\left(1 + \frac{SS_{\mathrm{H}}}{SS_{\mathrm{E}}}\right),$$

where

$$\frac{S(\tilde{\beta}) - S(\hat{\beta})}{S(\hat{\beta})} = \frac{SS_{\mathrm{H}}}{SS_{\mathrm{E}}} = \frac{q}{n-d}F_{\mathrm{H}}(y).$$

The statistic $F_{\mathrm{H}}(y)$ has the F-ratio distribution with q numerator and $n-d$ denominator degrees of freedom.

In the nonlinear regression model let β^* denote the "true" value of the parameter vector satisfying $g(\beta^*) = 0$. Now if the approximation

$$g(\beta) \approx g(\beta^*) + \frac{\partial g(\beta^*)}{\partial \beta^{\mathrm{T}}}(\beta - \beta^*)$$

is valid then the condition $g(\beta) = 0$ is approximately linear hypothesis on β. Also if the regression surface is well approximated by the tangent "plane" at β^* it follows that the distribution of

$$\frac{n-d}{q}\frac{S(\tilde{\beta}) - S(\hat{\beta})}{S(\hat{\beta})}$$

is approximately F-ratio distribution with q numerator and $n-d$ denominator degrees of freedom. That is the distribution of the likelihood ratio statistic

$$2\{\ell(\hat{\beta}, \hat{\sigma}; y_{\text{obs}}) - \ell(\tilde{\beta}, \tilde{\sigma}; y_{\text{obs}})\} = n \ln \left(1 + \frac{S(\tilde{\beta}) - S(\hat{\beta})}{S(\hat{\beta})}\right)$$

is approximately equal to the distribution of

$$n \ln \left(1 + \frac{q}{n-d} F(y)\right),$$

where $F(y)$ has the F-ratio distribution with q numerator and $n - d$ denominator degrees of freedom.

4.4   Profile likelihood-based confidence region

In a nonlinear regression model the approximate $(1-\alpha)$-level profile likelihood-based confidence region for the regression parameters has the form

$$\left\{\beta : \ S(\beta) \leq n \left(1 + \frac{d}{n-d} F_{1-\alpha}[d, n-d]\right) \hat{\sigma}^2\right\}, \qquad (4.15)$$

where $F_{1-\alpha}[q, n-d]$ is $(1-\alpha)^{\text{th}}$ quantile of F-distribution with q numerator and $n - d$ denominator degrees of freedom. This follows from the results of Section 4.3 by taking $g(\beta) = \beta$.

Example 4.3 Puromycin data (cont'd)

Figure 4.2 gives a plot of approximate 95%-level confidence region for regression parameters. From the confidence region one can infer that parameters are highly dependent and that statistical evidence supports values from 195 to 235 for β_1 and those from 0.044 to 0.094 for β_2. Separate confidence intervals for them will be calculated in the next section. ▯

4.5   Profile likelihood-based confidence interval

The approximate $(1 - \alpha)$-level profile likelihood-based confidence interval for the real-valued function $\psi = g(\beta)$ has the form

$$\left\{\psi : \ S(\tilde{\beta}_\psi) \leq n \left(1 + \frac{1}{n-d} F_{1-\alpha}[1, n-d]\right) \hat{\sigma}^2\right\}, \qquad (4.16)$$

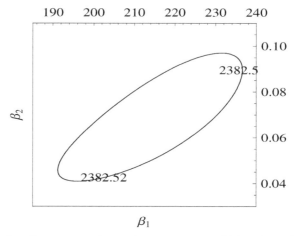

FIGURE 4.2: Puromycin data: Approximate 95%-level confidence region for regression parameters.

where $\tilde{\beta}_\psi$ is the MLE of the regression parameters under the hypothesis $H : g(\beta) = \psi$. In terms of log-likelihood function it has the form

$$\left\{ \psi : \ell(\tilde{\beta}_\psi, \tilde{\sigma}_\psi) \geq \ell(\hat{\beta}, \hat{\sigma}) - \frac{n}{2} \ln \left(1 + \frac{1}{n-d} F_{1-\alpha}[1, n-d] \right) \right\}. \qquad (4.17)$$

***Example 4.4* Puromycin data (cont'd)**

Table 4.2 contains MLEs and approximate 0.95-level profile likelihood-based confidence intervals for parameters and two interest functions. The first interest function is the mean value of the regression function at concentration 0.3, that is,

$$\psi_1 = \frac{0.3\beta_1}{0.3 + \beta_2},$$

and the second value of concentration at which the mean of the regression function takes the value 100 is

$$\psi_2 = \frac{100\beta_2}{\beta_1 - 100}.$$

In Table 4.3 corresponding intervals calculated from F-distribution with 1 numerator and 10 denominator degrees of freedom are shown. Figure 4.3 gives plots of profile likelihood functions of all considered interest functions. In these plots the indicated confidence intervals are based on χ^2-distribution with 1 degree of freedom.   ⊓

TABLE 4.2: Puromycin data: Estimates and approximate 0.95-level confidence intervals of interest functions.

Parameter function	Estimate	Lower limit	Upper limit
β_1	212.7	199.2	227.1
β_2	0.0641	0.0489	0.0830
σ	9.981	7.012	15.832
ψ_1	175.2	168.0	182.4
ψ_2	0.0570	0.0470	0.0681

TABLE 4.3: Puromycin data: Estimates and approximate 0.95-level confidence intervals based on F[1,10]distribution.

Parameter function	Estimate	Lower limit	Upper limit
β_1	212.7	197.3	229.3
β_2	0.0641	0.0469	0.0862
ψ_1	175.2	167.0	183.5
ψ_2	0.0570	0.0457	0.0699

Example 4.5 **Sulfisoxazole data (cont'd)**

The response concentration is modeled by a two-term pharmacokinetic model were the regression function has the form

$$f(t, \gamma_1, \lambda_1, \gamma_2, \lambda_2) = \gamma_1 e^{-\lambda_1 t} + \gamma_2 e^{-\lambda_2 t}.$$

Assuming that $\lambda_1 > \lambda_2$ the second term dominates for larger values of time and

$$\ln(y) \approx \ln(\gamma_2) - \lambda_2 t.$$

Thus, the following model matrix

$$\mathcal{L}X1 = \begin{pmatrix} 1 & -3. \\ 1 & -4. \\ 1 & -6. \\ 1 & -12. \\ 1 & -24. \\ 1 & -48. \end{pmatrix}$$

formed from the last six observations and the linear model

$$\mathcal{L}\mathcal{M}1 = \text{LinearRegressionModel}[\mathcal{L}X1]$$

for natural logarithms of the last six concentrations with MLE (5.051,0.153,0.0684) give starting values $e^{5.0507} = 156.131$ and 0.1528 for γ_2 and λ_2, respectively.

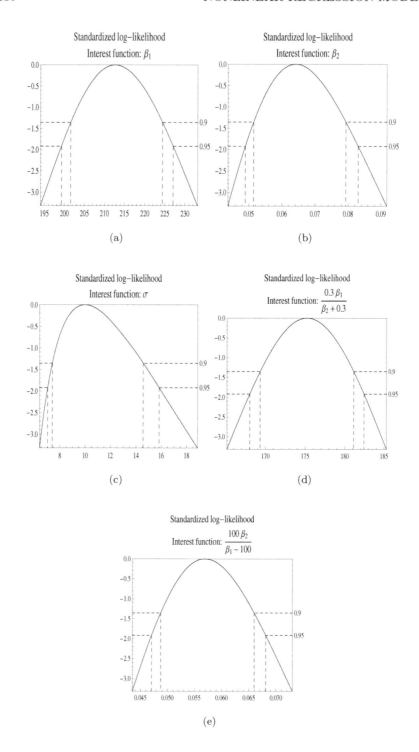

FIGURE 4.3:   Puromycin data: Plots of log-likelihoods of interest functions
(a) β_1, (b) β_2, (c) σ, (d) ψ_1, and (e) ψ_2.

Furthermore, the following model matrix

$$\mathcal{L}X2 = \begin{pmatrix} 1 & -0.25 \\ 1 & -0.50 \\ 1 & -0.75 \\ 1 & -1.00 \\ 1 & -1.50 \\ 1 & -2.00 \end{pmatrix}$$

and linear model

$$\mathcal{L}\mathcal{M}2 = \text{LinearRegressionModel}[\mathcal{L}X2]$$

for logarithms of the first six residuals calculated from the nonlinear regression model with MLE (4.556,1.323,0.0951) give starting values $e^{4.556} = 95.204$ and 1.323 for γ_1 and λ_1, respectively.

Finally, the model matrix

$$\mathcal{L}X3 = \begin{pmatrix} 1 \\ 1 \\ 1 \\ \vdots \\ 1 \\ 1 \\ 1 \end{pmatrix}$$

and linear model

$$\mathcal{L}\mathcal{M}3 = \text{LinearRegressionModel}[\mathcal{L}X3]$$

for all 12 residuals from the nonlinear regression model with MLE (-0.1787,2.316) give a starting value of 2.316 for the standard deviation.

The model matrix for the nonlinear regression model is

$$\mathcal{N}X = \begin{pmatrix} 0.25 \\ 0.50 \\ 0.75 \\ \vdots \\ 12.00 \\ 24.00 \\ 48.00 \end{pmatrix}$$

and the two-term pharmacokinetic model is denoted by

$$\mathcal{N}\mathcal{M} = \text{NonlinearRegressionModel}[\mathcal{N}X, \gamma_1 e^{-\lambda_1 t} + \gamma_2 e^{-\lambda_2 t}, \{t\}, \{\gamma_1, \lambda_1, \gamma_2, \lambda_2\}].$$

TABLE 4.4: Sulfisoxazole data: Estimates and approximate
0.95-level confidence intervals of interest functions.

Parameter function	Estimate	Lower limit	Upper limit
γ_1	81.24	70.78	91.86
λ_1	1.306	1.023	1.689
γ_2	162.6	150.2	173.4
λ_2	0.1618	0.1468	0.1764
σ	1.693	1.189	2.685
ψ_1	1067.2	1022.8	1116.4
ψ_2	424.1	417.5	430.6

TABLE 4.5: Sulfisoxazole data: Estimates and approximate
0.95-level confidence intervals based on F[1,8]distribution.

Parameter function	Estimate	Lower limit	Upper limit
γ_1	81.24	67.35	95.43
λ_1	1.306	0.945	1.846
γ_2	162.6	145.7	176.7
λ_2	0.1618	0.1418	0.1811
ψ_1	1067.2	1009.3	1133.8
ψ_2	424.1	415.3	432.8

The observed log-likelihood function is

$$-\frac{(215.6 - \gamma_1 e^{-0.25\lambda_1} - \gamma_2 e^{-0.25\lambda_2})^2}{2\sigma^2} - \cdots - \frac{(0.1 - \gamma_1 e^{-48.\lambda_1} - \gamma_2 e^{-48.\lambda_2})^2}{2\sigma^2}$$
$$-6\ln(2\pi) - 12\ln(\sigma).$$

Table 4.4 contains MLEs and approximate 0.95-level profile likelihood-based
confidence intervals of the regression parameters, standard deviation, and two
interest functions, namely, area under mean value curve (AUC) over the whole
positive axis

$$\psi_1 = \int_0^\infty (\gamma_1 e^{-\lambda t} + \gamma_2 e^{-\lambda_2 t})dt = \frac{\gamma_1}{\lambda_1} + \frac{\gamma_2}{\lambda_2}$$

and the same over time interval $(1,5)$

$$\psi_2 = \int_1^5 (\gamma_1 e^{-\lambda_1 t} + \gamma_2 e^{-\lambda_2 t})dt = \frac{\gamma_1}{\lambda_1}(e^{-\lambda_1} - e^{-5\lambda_1}) + \frac{\gamma_2}{\lambda_2}(e^{-\lambda_2} - e^{-5\lambda_2}).$$

In Table 4.5 the corresponding intervals calculated from F-distribution with 1
numerator and 8 denominator degrees of freedom are shown. Figure 4.4 gives
plots of log-likelihood functions of all considered interest functions. In these
plots the indicated confidence intervals are based on χ^2-distribution with 1
degree of freedom.   ▯

4.6   LRT for a hypothesis on finite set of functions

Consider the hypothesis H : $g(\beta) = 0$, where g is a q-dimensional vector valued function. According to Section 4.4, the likelihood ratio test statistic has the form

$$2\{\ell(\hat{\beta}, \hat{\sigma}; y_{\text{obs}}) - \ell(\tilde{\beta}, \tilde{\sigma}; y_{\text{obs}})\} = n \ln \left(\frac{S(\tilde{\beta})}{S(\hat{\beta})} \right), \qquad (4.18)$$

where $\tilde{\beta}$ is the maximum likelihood estimate of the regression parameter under the hypothesis H : $g(\beta) = 0$. Under the hypothesis the LRT statistic has approximately the same distribution as

$$n \ln \left(1 + \frac{q}{n - d} F(y) \right),$$

where $F(y)$ has the F-distribution with q numerator and $n - d$ denominator degrees of freedom.

Example 4.6 Puromycin data (cont'd)

Let us consider both treated and untreated groups. The values of velocity of enzymatic reaction form the response and the model matrix is

$$\mathcal{N}X = \begin{pmatrix} 0.02\ 1\ 0 \\ 0.02\ 1\ 0 \\ 0.02\ 0\ 1 \\ \vdots\ \vdots\ \vdots \\ 1.10\ 1\ 0 \\ 1.10\ 1\ 0 \\ 1.10\ 0\ 1 \end{pmatrix}$$

and the regression function is

$$f(x_1, x_2, x_3, \beta_1, \beta_2, \beta_3, \beta_4) = \frac{(x_2\beta_1 + x_3\beta_3)x_1}{x_1 + x_2\beta_2 + x_3\beta_4}.$$

The nonlinear regression model is

$$\mathcal{N}M = \text{NonlinearRegressionModel}[\mathcal{N}X, \frac{(x_2\beta_1 + x_3\beta_3)x_1}{x_1 + x_2\beta_2 + x_3\beta_4},$$
$$\{x_1, x_2, x_3\}, \{\beta_1, \beta_2, \beta_3, \beta_4\}].$$

By inspection we find that in addition to the starting values $\beta_1 = 207$ and $\beta_2 = 0.06$ the corresponding values for the untreated group are $\beta_3 = 160$ and

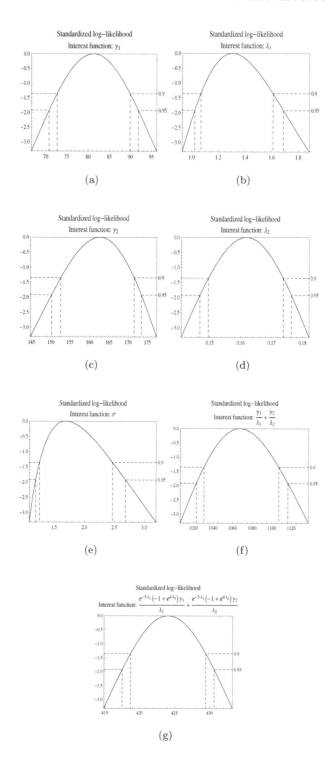

FIGURE 4.4: Sulfisoxazole data: Plots of log-likelihoods of interest functions (a) γ_1, (b) λ_1, (c) γ_2, (d) λ_2, (e) σ, (f) ψ_1, and (g) ψ_2.

$\beta_4 = 0.06$. As the starting value of the standard deviation, let us use that of the treated group.

MLEs of the vector of regression parameters and standard deviation are (212.62,0.0641,160.26,0.0477,9.453). The approximate observed significance level of the statistical hypothesis H : $\beta_2 = \beta_4$ is 0.2056. □

Example 4.7 Sulfisoxazole data (cont'd)

The significance level of the likelihood ratio test of the hypothesis H : $\gamma_2 = 0, \lambda_2 = 0$ is 4.427×10^{-6}. □

4.7 Model checking

Assume that the mean of the ith component of the response vector is some nonlinear function of the regression parameters

$$\mu_i = \eta_i = \eta_i(\beta).$$

Then the deviance residual of the ith component of the response vector is

$$d_i = \frac{y_i - \hat{\mu}_i}{\sigma},$$

where

$$\hat{\mu}_i = \eta_i(\hat{\beta}).$$

Because in the nonlinear regression model

$$W(\beta) = \text{diag}\left(\frac{1}{\sigma^2}, \ldots, \frac{1}{\sigma^2}\right),$$

the hat matrix is equal to

$$H(\hat{\beta}) = X(\hat{\beta})(X(\hat{\beta})^{\text{T}}X(\hat{\beta}))^{-1}X(\hat{\beta})^{\text{T}},$$

where

$$X(\beta) = \frac{\eta(\beta)}{\partial\beta^{\text{T}}} = \left(\frac{\eta_i(\beta)}{\partial\beta_j}\right).$$

The standardized deviance residuals have the form

$$r_{Di} = \frac{y_i - \hat{\mu}_i}{\sigma\sqrt{1 - h_{ii}}}. \tag{4.19}$$

The standardized Pearson residual is

$$r_{Pi} = \frac{u_i(\hat{\beta})}{\sqrt{w_i(\hat{\beta})(1 - h_{ii})}} = \frac{y_i - \hat{\mu}_i}{\sigma\sqrt{1 - h_{ii}}} = r_{Di}. \tag{4.20}$$

Thus, in the nonlinear regression model all three residuals are equal, that is,

$$r_{Di} = r_{Pi} = r_i^*.$$

Because σ is unknown the actual residuals will be obtained by replacing σ with its estimate, that is, the MLE $\hat{\sigma}$ and so one gets

$$r_i^* = r_{Di} = r_{Pi} = \frac{y_i - \hat{\mu}_i}{\hat{\sigma}\sqrt{(1 - h_{ii})}} \quad \forall i = 1, \ldots, n.$$

Asymptotically these residuals have marginal standard normal distribution, but in this case one can say more. Namely, using unbiased estimate s instead of $\hat{\sigma}$, that is, considering

$$r_i^* = r_{Di} = r_{Pi} = \frac{y_i - \hat{\mu}_i}{s\sqrt{(1 - h_{ii})}} \quad \forall i = 1, \ldots, n \qquad (4.21)$$

one gets *studentized residuals* the marginal distribution of which is approximately Student t-distribution with $n - d$ degrees of freedom.

Example 4.8 Puromycin data (cont'd)

Figure 4.5 gives a quantile-quantile plot along with plots of standardized residuals against concentration and fitted values.   ⬜

Example 4.9 Sulfisoxazole data (cont'd)

Figure 4.6 gives a quantile-quantile plot along with plots of standardized residuals against time and fitted values.   ⬜

4.8   Bibliographic notes

Bates and Watts (1988) is an excellent book on nonlinear regression analysis.

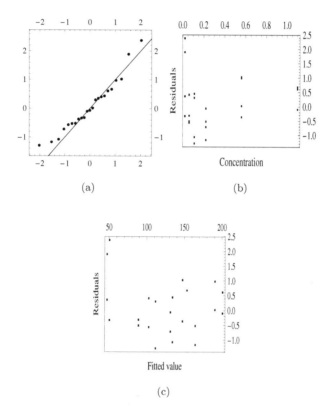

FIGURE 4.5: Puromycin data: (a) QQ-plot and plots of residuals against (b) concentration and (c) fitted values.

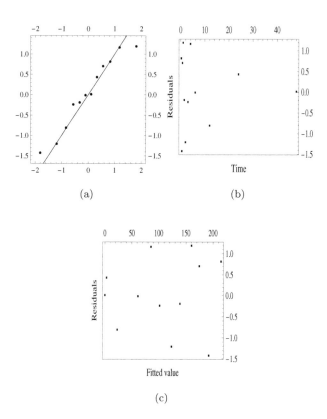

(a)

(b)

(c)

FIGURE 4.6:   Sulfisoxazole data: (a) QQ-plot and plots of residuals against (b) time and (c) fitted values.

Chapter 5

Generalized Linear Model

5.1 Definition of generalized linear model

5.1.1 Distribution, linear predictor, and link function

The generalized linear model (GLIM) was defined in Section 2.3 as a univariate and one-parameter GRM with statistically independent components of the response vector such that transformed means of the components depend linearly on the regression coefficients. "Classically" this is presented using the following definitions.

GLIM consists of the following parts:

1. statistically independent responses $y_1, \ldots, y_n$, whose point probability or density functions have the form

$$p(y_i; \theta_i, \phi) = e^{\frac{y_i \theta_i - b(\theta_i)}{a_i(\phi)} - c(y_i, \phi)} \ \forall i = 1, \ldots, n \qquad (5.1)$$

and their means being $\mu_1, \ldots, \mu_n$,

2. *linear predictor*

$$\eta_i = \sum_{j=1}^{d} x_{ij} \beta_j = x_i \beta \ \forall i = 1, \ldots, n \qquad (5.2)$$

where the x_{ij}s $(j = 1, \ldots, d)$ are values of regressors for the response y_i and form the ith row x_i of the model matrix X and $\beta_1, \ldots, \beta_d$ are the regression coefficients, and

3. known *link function* g, which gives the relation between the linear predictor and the mean of the response, that is,

$$\eta_i = g(\mu_i) \ \forall i = 1, \ldots, n. \qquad (5.3)$$

The link function g, which satisfies $\theta_i = g(\mu_i) = \eta_i = x_i \beta$, is called the *canonical link function*.

In Equation (5.1) the model function $p(y_i; \theta_i, \phi)$ is defined on some given subset of real line, θ_i is an unknown real number, ϕ is an unknown/known positive real number, and a_i, b, and c are given functions of their arguments. In the following section exact forms of these components are given for five considered distributions, namely, normal, Poisson, binomial, gamma, and inverse Gaussian distributions. If ϕ is known, the distribution of the response is 1-*parameter exponential family*.

The log-likelihood function connected to the response y_i is

$$\ell(\theta_i, \phi; y_i) = \frac{y_i\theta_i - b(\theta_i)}{a_i(\phi)} - c(y_i, \phi) \tag{5.4}$$

and the score function is

$$\frac{\partial \ell(\theta_i, \phi; y_i)}{\partial \theta_i} = \frac{y_i - b'(\theta_i)}{a_i(\phi)}. \tag{5.5}$$

From the properties of the score function it follows that

$$E\left(\frac{\partial \ell}{\partial \theta_i}\right) = \frac{E(y_i) - b'(\theta_i)}{a_i(\phi)} = 0$$

or

$$\mu_i = \mu_i(\theta_i) = E(y_i; \theta_i, \phi) = b'(\theta_i) \tag{5.6}$$

and

$$\frac{\text{var}(y_i)}{a_i(\phi)^2} = \text{var}\left(\frac{\partial \ell}{\partial \theta_i}\right)$$

$$= E\left(-\frac{\partial^2 \ell}{\partial \theta_i^2}\right)$$

$$= \frac{b''(\theta_i)}{a_i(\phi)}$$

or

$$\text{var}(y_i; \theta_i, \phi) = a_i(\phi)b''(\theta_i) > 0 \ \forall i = 1, \ldots, n, \tag{5.7}$$

where $a_i(\phi)$ are positive for all $i = 1, \ldots, n$.

Usually

$$a_i(\phi) = \frac{\phi}{w_i} \ \forall i = 1, \ldots, n,$$

where the w_is are *prior weights* of the observations. From the expression of variance follows that $b'(\theta)$ is a strictly increasing function of θ and there exists a strictly increasing function $h = (b')^{-1}$ such that

$$\theta_i = h(\mu_i) \ \forall i = 1, \ldots, n.$$

Thus, the variance of the response y_i has the form

$$\text{var}(y_i; \theta_i, \phi) = a_i(\phi)V(\mu_i) \ \forall i = 1, \ldots, n, \tag{5.8}$$

where

$$V(\mu) = b''(h(\mu)) = b''((b')^{-1}(\mu)) \qquad (5.9)$$

is the *variance function*. Thus, the variance of the response can be decomposed into two parts a and V such that V depends only on the mean and a depends only on ϕ (and on the known prior weights). Accordingly ϕ is known as the *dispersion parameter* or *scale parameter*.

In the case of the canonical link function the variance function is

$$V(\mu) = b''(h(\mu)) = \frac{1}{g'(\mu)}$$

and the log-likelihood function has the form

$$\ell(\beta, \phi; y) = \sum_i \left\{ \frac{y_i x_i \beta - b(x_i \beta)}{a_i(\phi)} - c(y_i, \phi) \right\}$$

$$= \sum_j \left(\sum_i \frac{x_{ij} y_i}{a_i(\phi)} \right) \beta_j - \sum_i \left(\frac{b(x_i \beta)}{a_i(\phi)} + c(y_i, \phi) \right). \qquad (5.10)$$

When ϕ is known the statistical model is a d-dimensional exponential family and the statistics

$$\sum_i \frac{x_{ij} y_i}{a_i(\phi)} \quad \forall j = 1, \ldots, d,$$

that is, the vector

$$X^T W y,$$

where

$$W = \text{diag} \left(\frac{1}{a_1(\phi)}, \ldots, \frac{1}{a_n(\phi)} \right),$$

is *minimal sufficient statistic*.

5.1.2   Examples of distributions generating generalized linear models

Example 5.1 NormalModel[μ, σ]

Table 5.1 gives the definition of the normal distribution in terms of the concepts of the previous section. The model consisting of normal distribution with unknown mean μ and unknown standard deviation σ is denoted by

$$\mathcal{M} = \text{NormalModel}[\mu, \sigma].$$

The density function $p(y; \mu, \sigma)$ of the normal distribution written in conventional form is

$$\frac{e^{-\frac{(y-\mu)^2}{2\sigma^2}}}{\sqrt{2\pi}\sigma} \qquad (5.11)$$

TABLE 5.1:   NormalModel.

y	ϕ	$b(\theta)$	$c(y;\phi)$	$\mu(\theta)$	$\theta(\mu)$	$V(\mu)$
$\mathbb{R}$	σ^2	$\frac{\theta^2}{2}$	$-\frac{1}{2}(\frac{y^2}{\phi}+\ln(2\pi\phi))$	θ	μ	1

for all $y \in \mathbb{R}$. Figure 5.1 contains plots of density function for selected values of mean and standard deviation.

Canonical link function is

$$\theta_i = \theta(\mu_i) = \mu_i = \eta_i = x_i\beta, \tag{5.12}$$

that is,

$$y_i = \mu_i + \epsilon_i = x_i\beta + \epsilon_i \ \forall i = 1,\ldots,n.$$

So the general linear model (GLM) is a GLIM, whose distribution is normal distribution and link function is canonical link function. GLM has already been considered in Chapter 3.   □

Example 5.2 PoissonModel[μ]

Table 5.2 gives the definition of Poisson distribution in terms of the concepts of the previous section.

TABLE 5.2:   PoissonModel.

y	ϕ	$b(\theta)$	$c(y;\phi)$	$\mu(\theta)$	$\theta(\mu)$	$V(\mu)$
$\{0,1,2,\ldots\}$	1	e^θ	$-\ln(y!)$	e^θ	$\ln(\mu)$	μ

The model consisting of Poisson distribution with unknown mean μ is denoted by

$$\mathcal{M} = \text{PoissonModel}[\mu].$$

The point probability function $p(y;\mu)$ of the Poisson distribution written in conventional form is

$$\frac{\mu^y}{y!}e^{-\mu} \tag{5.13}$$

for all $y \in \{0,1,2,\ldots\}$. Figure 5.2 contains plots of point probability function for selected values of mean.

The canonical link function is

$$\theta_i = \theta(\mu_i) = \ln(\mu_i) = \eta_i = x_i\beta, \tag{5.14}$$

that is,

$$p(y_i;\beta) = \frac{e^{y_i x_i\beta}}{y_i!}e^{-e^{x_i\beta}}.$$

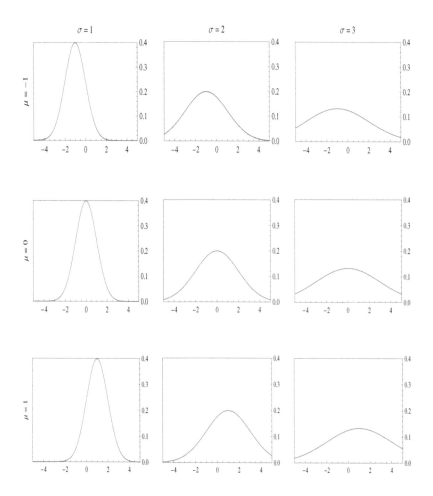

FIGURE 5.1: NormalModel: Graphs of densities.

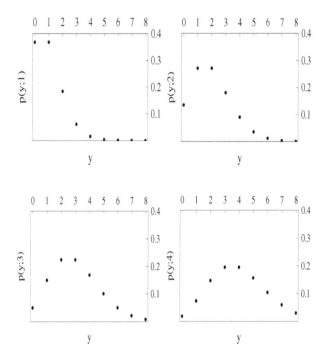

FIGURE 5.2:   PoissonModel: Graphs of point probability functions.

GLIM with Poisson distribution is considered in more detail in Chapter 7.
⬚

Example 5.3 **BinomialModel**[m, μ]

Table 5.3 gives the definition of a binomial distribution in terms of the concepts of the previous section.

In Table 5.3 the assumption that my has binomial distribution gives previous definitions. In the following, however, the conventional assumption, that is, y has binomial distribution, is always used.

The model consisting of binomial distribution with 10 trials and unknown

TABLE 5.3:   BinomialModel.

y	ϕ	$b(\theta)$	$c(y;\phi)$	$\mu(\theta)$	$\theta(\mu)$	$V(\mu)$
$\{0, \frac{1}{m}, \frac{2}{m}, \ldots, 1\}$	$\frac{1}{m}$	$\ln(1 + e^{\theta})$	$\ln\binom{m}{my}$	$\frac{e^{\theta}}{1+e^{\theta}}$	$\ln\left(\frac{\mu}{1-\mu}\right)$	$\mu(1 - \mu)$

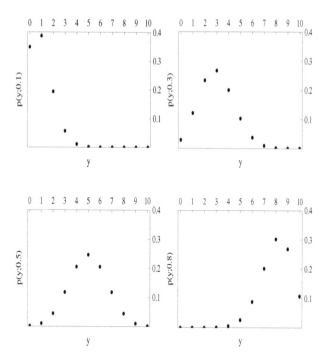

FIGURE 5.3: BinomialModel: Graphs of point probability functions.

success probability ω is denoted by

$$\mathcal{M} = \text{BinomialModel}[10, \omega].$$

The point probability function $p(y; \omega)$ of the binomial distribution with 10 trials written in conventional form is

$$\binom{10}{y} \omega^y (1 - \omega)^{10-y} \tag{5.15}$$

for all $y \in \{0, 1, 2, \ldots, 10\}$. Figure 5.3 contains plots of point probability function for selected values of success probability.

The canonical link function is

$$\theta_i = \theta(\mu_i) = \ln\left(\frac{\mu_i}{m_i - \mu_i}\right) = \ln\left(\frac{\omega_i}{1 - \omega_i}\right) = \eta_i = x_i\beta, \tag{5.16}$$

that is,

$$p(y_i; \beta) = \binom{m_i}{y_i}\left(\frac{e^{x_i\beta}}{1 + e^{x_i\beta}}\right)^{y_i}\left(\frac{1}{1 + e^{x_i\beta}}\right)^{m_i - y_i}$$

$$= \binom{m_i}{y_i}\frac{e^{y_i x_i\beta}}{(1 + e^{x_i\beta})^{m_i}}.$$

TABLE 5.4: GammaModel.

y	ϕ	$b(\theta)$	$c(y;\phi)$	$\mu(\theta)$	$\theta(\mu)$	$V(\mu)$
$(0,\infty)$	$\frac{1}{\nu}$	$-\ln(-\theta)$	$\nu\ln(\nu y)-\ln(y)-\ln(\Gamma(y))$	$-\frac{1}{\theta}$	$-\frac{1}{\mu}$	μ^2

GLIM with binomial distribution is considered in more detail in Chapter 6. In that chapter also cases with parametrized and nonlinear link functions will be presented. ⬜

Example 5.4 GammaModel[ν, μ]

Table 5.4 gives the definition of gamma distribution in terms of the concepts of the previous section.

The model consisting of gamma distribution with unknown shape ν and unknown mean μ is denoted by

$$\mathcal{M} = \text{GammaModel}[\nu,\mu].$$

The density function $p(y;\nu,\mu)$ of the gamma distribution written in conventional form is

$$\frac{(\frac{\nu}{\mu})^\nu y^{\nu-1} e^{-\frac{y\nu}{\mu}}}{\Gamma(\nu)} \tag{5.17}$$

for all $y \in (0,\infty)$. Figure 5.4 contains plots of density function for selected values of shape and mean.

The canonical link function is

$$\theta_i = \theta(\mu_i) = -\frac{1}{\mu_i} = \eta_i = x_i\beta. \tag{5.18}$$

GLIM with gamma distribution is considered in more detail in Section 10.2. ⬜

Example 5.5 InverseGaussianModel[μ, τ]

Table 5.5 gives the definition of inverse Gaussian distribution in terms of the concepts of the previous section.

TABLE 5.5: InverseGaussianModel.

y	ϕ	$b(\theta)$	$c(y;\phi)$	$\mu(\theta)$	$\theta(\mu)$	$V(\mu)$
$(0,\infty)$	τ	$-\sqrt{-2\theta}$	$-\frac{1}{2}\left\{\ln(2\pi\phi y^3)+\frac{1}{\phi y}\right\}$	$\frac{1}{\sqrt{-2\theta}}$	$\frac{1}{\mu^2}$	μ^3

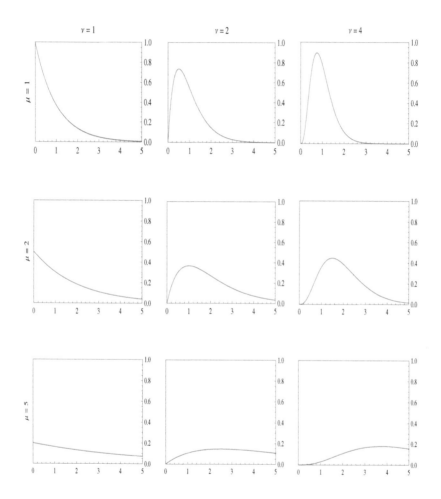

FIGURE 5.4:   GammaModel: Graphs of densities.

The model consisting of inverse Gaussian distribution with unknown mean μ and unknown variance τ is denoted by

$$\mathcal{M} = \text{InverseGaussianModel}[\mu, \tau].$$

The density function $p(y; \mu, \tau)$ of the inverse Gaussian distribution written in conventional form is

$$\frac{\sqrt{\frac{\mu^3}{y^3}} e^{-\frac{(y-\mu)^2 \mu}{2y\tau}}}{\sqrt{2\pi}\sqrt{\tau}} \tag{5.19}$$

for all $y \in (0, \infty)$. Figure 5.5 contains plots of the density function for selected values of shape and mean.

The canonical link function is

$$\theta_i = \theta(\mu_i) = \frac{1}{\mu_i^2} = \eta_i = x_i \beta. \tag{5.20}$$

☐

5.2 MLE of regression coefficients

5.2.1 MLE

It is instructive to consider MLE in detail even though the more general case has already appeared in Section 2.5. The likelihood equations have the form

$$\frac{\partial \ell}{\partial \beta_j} = \sum_{i=1}^{n} \frac{\partial \ell(\theta_i, \phi; y_i)}{\partial \beta_j}$$

$$= \sum_{i=1}^{n} \frac{\partial \ell(\theta_i, \phi; y_i)}{\partial \theta_i} \frac{d\theta_i}{d\mu_i} \frac{d\mu_i}{d\eta_i} \frac{\partial \eta_i}{\partial \beta_j} = 0 \; \forall j = 1, \ldots, d.$$

Because

$$\frac{\partial \ell(\theta_i, \phi; y_i)}{\partial \theta_i} = \frac{\partial}{\partial \theta_i} \left(\frac{y_i \theta_i - b(\theta_i)}{a_i(\phi)} \right) = \frac{y_i - b'(\theta_i)}{a_i(\phi)} = \frac{y_i - \mu_i}{a_i(\phi)},$$

$$\frac{d\theta_i}{d\mu_i} = \frac{1}{\frac{d\mu_i}{d\theta_i}} = \frac{1}{\frac{db'(\theta_i)}{d\theta_i}} = \frac{1}{b''(\theta_i)} = \frac{1}{V(\mu_i)},$$

$$\frac{d\mu_i}{d\eta_i} = \frac{1}{\frac{d\eta_i}{d\mu_i}} = \frac{1}{g'(\mu_i)},$$

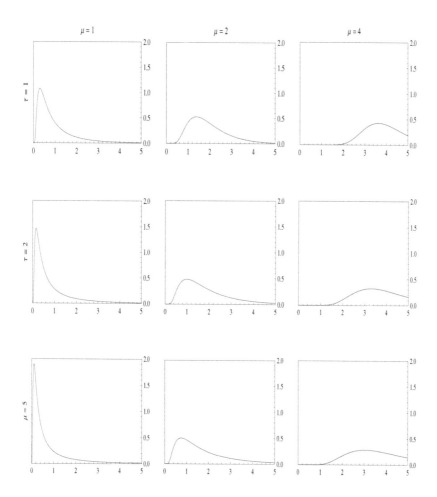

FIGURE 5.5: InverseGaussianModel: Graphs of densities.

and

$$\frac{\partial \eta_i}{\partial \beta_j} = x_{ij},$$

the likelihood equations can be written in the following form:

$$\frac{\partial \ell}{\partial \beta_j} = \sum_{i=1}^{n} \frac{y_i - \mu_i}{a_i(\phi)} \frac{1}{V(\mu_i)} \frac{1}{g'(\mu_i)} x_{ij} = 0 \ \forall j = 1, \ldots, d.$$

Because in the case of the canonical link function $V(\mu)g'(\mu) = 1$, the likelihood equations are in this case

$$\frac{\partial \ell}{\partial \beta_j} = \sum_{i=1}^{n} \frac{y_i - \mu_i}{a_i(\phi)} x_{ij} = 0 \ \forall j = 1, \ldots, d$$

or

$$E(X^{\mathrm{T}} W y) = X^{\mathrm{T}} W y,$$

that is, MLE is that value of the parameter vector for which the expected value of the minimal sufficient statistic is equal to its observed value.

When $a_i(\phi) = \frac{\phi}{w_i}$, the likelihood equations take the form

$$\frac{\partial \ell}{\partial \beta_j} = \frac{1}{\phi} \sum_{i=1}^{n} (y_i - \mu_i) u_i g'(\mu_i) x_{ij} = 0 \ \forall j = 1, \ldots, d,$$

or

$$\sum_{i=1}^{n} (y_i - \mu_i) u_i g'(\mu_i) x_{ij} = 0 \ \forall j = 1, \ldots, d,$$

where

$$u_i = \frac{w_i}{V(\mu_i) g'(\mu_i)^2} \ \forall i = 1, \ldots, n.$$

5.2.2   Newton-Raphson and Fisher-scoring

Because $\mu_i = h(\eta_i) = h(x_i\beta) \ \forall i = 1, \ldots, n$, the likelihood equations form a set of nonlinear equations, the solution of which is the MLE of β. The *Newton-Raphson procedure* for the solution of the set of nonlinear equations $\frac{\partial \ell}{\partial \beta_j} = 0 \ \forall j = 1, \ldots, d$ proceeds according to the following iteration procedure:

$$\beta^{(r)} = \beta^{(r-1)} + \delta\beta^{(r-1)}$$

where

$$\delta\beta^{(r-1)} = \left(-\left(\frac{\partial^2 \ell}{\partial \beta \partial \beta^{\mathrm{T}}} \right)^{-1} \frac{\partial \ell}{\partial \beta} \right)_{\beta=\beta^{(r-1)}}.$$

In the *Fisher-scoring procedure* the second partial derivatives of the log-likelihood function will be replaced by their expectations with the result that sometimes the formulas simplify.

In the case of GLIM

$$E\left(\frac{\partial^2 \ell}{\partial \beta_j \partial \beta_k}\right) = -E\left(\frac{\partial \ell}{\partial \beta_j} \frac{\partial \ell}{\partial \beta_k}\right)$$

$$= -\frac{1}{\phi^2} E\left(\left(\sum_{i=1}^n (y_i - \mu_i) u_i g'(\mu_i) x_{ij}\right)\left(\sum_{m=1}^n (y_m - \mu_m) u_m g'(\mu_m) x_{mk}\right)\right).$$

When the components of the response vector are statistically independent,

$$E\left(\frac{\partial \ell}{\partial \beta_j} \frac{\partial \ell}{\partial \beta_k}\right) = \frac{1}{\phi^2} \sum_{i=1}^n E((y_i - \mu_i)^2 u_i^2 (g'(\mu_i))^2) x_{ij} x_{ik}$$

$$= \frac{1}{\phi^2} \sum_{i=1}^n \text{var}(y_i) u_i^2 (g'(\mu_i))^2 x_{ij} x_{ik},$$

and, because $\text{var}(y_i) = a_i(\phi) V(\mu_i)$, we get finally

$$E\left(\frac{\partial^2 \ell}{\partial \beta_j \partial \beta_k}\right) = -\frac{1}{\phi} \sum_{i=1}^n u_i x_{ij} x_{ik}.$$

Thus, for the Fisher-scoring procedure

$$\beta^{(r)} = \beta^{(r-1)} + \left(X^T U^{(r-1)} X\right)^{-1} X^T U^{(r-1)} \nu^{(r-1)},$$

where

$$U^{(r-1)} = \text{diag}\left(u_1^{(r-1)}, \ldots, u_n^{(r-1)}\right)$$

and

$$\nu_i^{(r-1)} = \left(y_i - \mu_i^{(r-1)}\right) g'\left(\mu_i^{(r-1)}\right) \; \forall i = 1, \ldots, n.$$

Finally,

$$X^T U^{(r-1)} X \beta^{(r)} = X^T U^{(r-1)} z^{(r-1)},$$

where

$$z_i^{(r-1)} = \eta_i^{(r-1)} + \left(y_i - \mu_i^{(r-1)}\right) g'\left(\mu_i^{(r-1)}\right) \; \forall i = 1, \ldots, n$$

and z is the adjusted response. The previous results are clearly the ones appearing in Section 2.5 in the case of GLIM.

5.3   Bibliographic notes

An elementary book on generalized linear regression models is Dobson (1990) and McCullagh and Nelder (1989) is the "classical" treatise on the subject.

Jørgensen (1997b) contains a discussion on distributions of generalized linear models. Thompson and Baker (1981) includes material on parametrized link functions.

The article by Nelder and Wedderburn (1972) introduced the theory of generalized linear models.

Chapter 6

Binomial and Logistic Regression Model

6.1   Data

Suppose that a binary response z, the two possible values of which are denoted by 0 and 1, has been observed from every statistical unit. "Failure" (0) and "success" (1) probabilities for the statistical unit j are denoted by

$$\Pr(z_j = 0) = 1 - \omega_j \text{ and } \Pr(z_j = 1) = \omega_j \ \forall j = 1, \ldots, N.$$

Usually success probability ω_j of the jth statistical unit is assumed to depend on values $(x_{j1}, \ldots, x_{jq})$ of some factors associated with the unit, that is,

$$\Pr(z_j = 1) = \omega_j = \omega(x_{j1}, \ldots, x_{jq}) \ \forall j = 1, \ldots, N,$$

where ω is a given function of the factors $x_1, \ldots, x_q$.

Let there be in data n different combinations $(x_{i1}, \ldots, x_{iq}) \ \forall i = 1, \ldots, n$ of values of the factors and let there be m_i statistical units with the combination $(x_{i1}, \ldots, x_{iq})$ of the factors, that is, m_i statistical units in the data have same combination $(x_{i1}, \ldots, x_{iq})$ of values of the factors. These units are said to form a *covariate class*.

When in the data the number of covariate classes is considerably smaller than the number of statistical units, it is practical to generate new variables $y_i \ \forall i = 1, \ldots, n$, where y_i denotes the total number of successes in the ith covariate class. Now the new responses y_i are statistically independent and

$$y_i \sim \text{BinomialModel}[m_i, \omega_i] \ \forall i = 1, \ldots, n, \tag{6.1}$$

where

$$\omega_i = \omega(x_{i1}, \ldots, x_{iq}) \ \forall i = 1, \ldots, n$$

is the success probability of the ith covariate class. In the process of aggregation of the data into the counts of successes in covariate classes information may be lost. The original or ungrouped data can be considered to be a special case, where

$$m_1 = m_2 = \cdots = m_N = 1.$$

Making a difference between grouped and ungrouped data is important for at least two reasons.

1. Certain methods of analysis, especially those relying on the normal approximation, which are applicable in the case of grouped data, are not applicable to ungrouped data.

2. The asymptotic approximations of grouped data may be based either on the property $m_i \to \infty \ \forall i = 1, \ldots, n$ or on the property $N \to \infty$. Only the latter property is applicable to ungrouped data.

Assume now that the function $w(x_1, \ldots, x_q)$ in addition to factors depends on d unknown common parameters $\beta_1, \ldots, \beta_d$, that is,

$$w_i(\beta_1, \ldots, \beta_d) = w(x_{i1}, \ldots, x_{iq}; \beta_1, \ldots, \beta_d) \ \forall i = 1, \ldots, n. \qquad (6.2)$$

Then the model is a GRM. We call it *binomial regression model* in general. This includes cases where success probabilities depend nonlinearly on regression parameters. If the function w has the form

$$w(x_{i1}, \ldots, x_{id}; \beta_1, \ldots, \beta_d) = h(x_{i1}\beta_1 + x_{i2}\beta_2 + \cdots + x_{id}\beta_d) \ \forall i = 1, \ldots, n \quad (6.3)$$

for some known function h, the model is a GLIM. The function h is inverse of the *link function*, that is, $g = h^{-1}$ is the link function. Finally when $h(z) = \frac{e^z}{1+e^z}$ the model is called *logistic regression model*.

Example 6.1 O-ring data

O-ring thermal distress data contains numbers of field-joint O-rings showing thermal distress out of six, for a launch at given temperature (°F) and pressure (pounds per square inch) (Dalal *et al.*, 1989). Columns of Table A.15 on page 262 in Appendix A show counts of thermal distress, temperature (°F), and pressure (psi). Figure 6.1 plots the proportion of thermal distress against temperature, pressure, and both.

Frequencies of thermal distress might be modeled as statistically independent having binomial distributions with six trials and success probabilities depending on the linear combination of temperature and pressure through logit-function, that is, logistic regression model. This model is denoted by the expression

$$\mathcal{M} = \text{LogisticRegressionModel}[X, 6],$$

where the first argument gives the model matrix, that is,

$$X = \begin{pmatrix} 1 & 66 & 50 \\ 1 & 70 & 50 \\ 1 & 69 & 50 \\ \vdots & \vdots & \vdots \\ 1 & 75 & 200 \\ 1 & 76 & 200 \\ 1 & 58 & 200 \end{pmatrix}$$

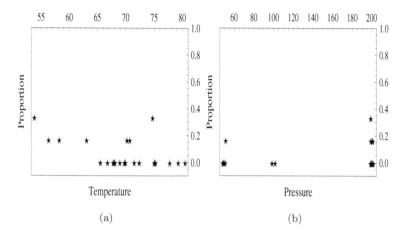

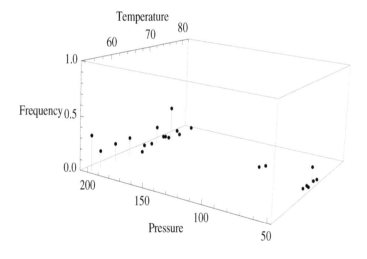

FIGURE 6.1:   O-ring data: Scatter plots of observed relative frequencies against (a) temperature, (b) pressure, (c) and both.

and the second argument gives the number of trials that is same for all components of the response vector.

Because proportion of distress does not seem to depend on pressure, Akaike's information criterion was used to compare models with or without it. AIC of the first model is 36.1059 and that of the latter is 35.6465. So only temperature is used to explain distress probability. ⬜

6.2 Binomial distribution

The moment generating function of BernoulliModel$[\omega]$ has the form

$$\mathrm{M}(\xi) = \mathrm{E}(e^{\xi y}) = 1 - \omega + \omega e^{\xi} \tag{6.4}$$

and its cumulant generating function is

$$\mathrm{K}(\xi) = \ln(\mathrm{M}_y(\xi)) = \ln(1 - \omega + \omega e^{\xi}). \tag{6.5}$$

Thus, the moment generating function of the BinomialModel$[m, \omega]$ is

$$(1 - \omega + \omega e^{\xi})^m. \tag{6.6}$$

Similarly the cumulant generating function is

$$m \ln(1 - \omega + \omega e^{\xi}). \tag{6.7}$$

The first four cumulants of the BinomialModel$[m, \omega]$ are

$$\kappa_1 = m\omega, \tag{6.8}$$
$$\kappa_2 = m\omega(1 - \omega), \tag{6.9}$$
$$\kappa_3 = m\omega(1 - \omega)(1 - 2\omega), \tag{6.10}$$
$$\kappa_4 = m\omega(1 - \omega)\{1 - 6\omega(1 - \omega)\}. \tag{6.11}$$

The cumulants of the standardized variable

$$z = \frac{y - m\omega}{\sqrt{m\omega(1 - \omega)}}$$

are seen from the Taylor series expansion of the cumulant generating function of z

$$\frac{\xi^2}{2} + \frac{(1 - 2\omega)\xi^3}{6\sqrt{m\omega(1 - \omega)}} + \frac{\left(1 - 6\omega + 6\omega^2\right)\xi^4}{24m\omega(1 - \omega)} +$$
$$\frac{\left(1 - 14\omega + 36\omega^2 - 24\omega^3\right)\xi^5}{120m^{3/2}\omega^{3/2}(1 - \omega)^{3/2}} +$$
$$\frac{\left(1 - 30\omega + 150\omega^2 - 240\omega^3 + 120\omega^4\right)\xi^6}{720m^2\omega^2(1 - \omega)^2} + O\left(\xi^7\right)$$

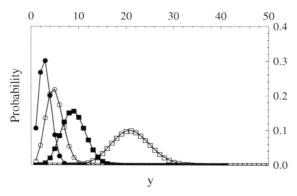

FIGURE 6.2:   Point probability function of binomial distribution with success probability 0.2 and number of trials 10 ($\bullet$), 20 ($\circ$), 40 ($\blacksquare$), and 100 ($\square$).

to be of order 0, 1, $O(m^{-1/2})$, $O(m^{-1})$, etc. When $r \geq 2$, r^{th} cumulant of z is of order $O(m^{1-r/2})$. When ω is fixed, the cumulants of z approach the cumulants 0, 1, 0, 0, ... of standardized normal distribution as $m \to \infty$. Because convergence of cumulants implies convergence in distribution, the approximate probabilities will be obtained from the formulae

$$\Pr(y \geq \tilde{y}) \approx 1 - \Phi(z^-)$$

$$\Pr(y \leq \tilde{y}) \approx \Phi(z^+),$$

where Φ is the cumulative distribution function of standardized normal distribution, $\tilde{y}$ is an integer, and

$$z^- = \frac{\tilde{y} - m\omega - \frac{1}{2}}{\sqrt{m\omega(1 - \omega)}} \text{ and } z^+ = \frac{\tilde{y} - m\omega + \frac{1}{2}}{\sqrt{m\omega(1 - \omega)}}.$$

Thus, the effect of the continuity correction $\mp \frac{1}{2}$ on the probabilities is of order $O(m^{-1/2})$ and so asymptotically negligible. In medium size samples, the effect of the continuity correction is remarkable and almost always improves the approximation.

Figure 6.2 contains graphs of point probability functions of binomial distributions with success probability 0.2 and numbers of trials 10, 20, 40, and 100.

6.3 Link functions

6.3.1 Unparametrized link functions

The following list gives the three common link functions and then generalizes the concept.

1. logit or logistic function

$$g_1(\omega) = \ln\left(\frac{\omega}{1-\omega}\right) = \text{logit}(\omega), \tag{6.12}$$

2. probit or inverse cumulative distribution function of standardized normal distribution

$$g_2(\omega) = \Phi^{-1}(\omega) = \text{probit}(\omega), \tag{6.13}$$

3. complementary log-log function

$$g_3(\omega) = \ln(-\ln(1-\omega)), \tag{6.14}$$

4. general link function

   In principle the inverse function of the cumulative distribution function F of any absolutely continuous distribution defined on the whole real line with positive density everywhere can be used as a link function, that is,

$$g_4(\omega) = F^{-1}(\omega). \tag{6.15}$$

Figure 6.3 compares the previous first three link functions.

Distributions corresponding to logit, probit, and complementary log-log links are standardized logistic, normal, and (negative of) extreme value distributions, respectively. Their cumulative distribution functions are $\frac{1}{1+e^{-z}}$, $\Phi(z)$, and $1 - e^{-e^z}$.

In the next example the cumulative distribution function of the Cauchy distribution with location 0 and scale 1 is used.

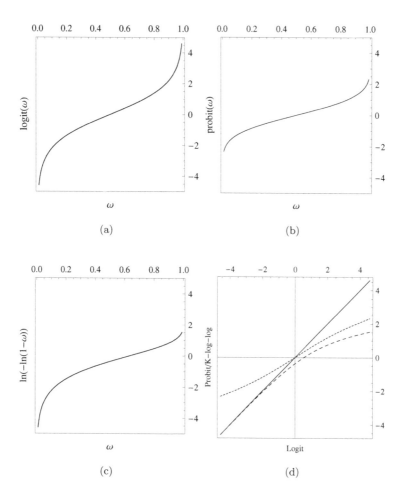

FIGURE 6.3: Binomial link functions: (a) logit, (b) probit, (c) complementary log-log, and (d) all three.

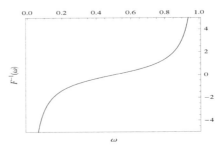

FIGURE 6.4: Inverse of cumulative distribution function of Cauchy-Model[0,1].

Example 6.2 Remission of cancer data (cont'd)

Assuming the labeling index affects linearly the log-odds we get the model matrix

$$X = \begin{pmatrix} 1 & 8 \\ 1 & 10 \\ 1 & 12 \\ \vdots & \vdots \\ 1 & 32 \\ 1 & 34 \\ 1 & 38 \end{pmatrix}.$$

Taking the inverse of the cumulative distribution function of the standard Cauchy distribution as the link function means that the inverse link function has the form

$$h(\eta) = g^{-1}(\eta) = \frac{\tan^{-1}(\eta)}{\pi} + \frac{1}{2} \ \forall \eta \in \mathbb{R}.$$

Figure 6.4 shows the link function.

The binomial regression model with this function as the inverse link function is denoted by the expression

$\mathcal{M} = \text{RegressionModel}[X,$
$\qquad \text{Distribution} \rightarrow \{\text{BinomialModel}[2, \omega], \ldots, \text{BinomialModel}[3, \omega]\},$
$\qquad \text{InverseLink} \rightarrow h],$

where the expressions

$\qquad \text{Distribution} \rightarrow \{\text{BinomialModel}[2, \omega], \ldots, \text{BinomialModel}[3, \omega]\}$

and

$$\text{InverseLink} \rightarrow h$$

denote so-called optional arguments whose default values are NormalModel[μ, σ] and identity function. The first of these tells that components of the response vector are assumed to be statistically independent observations from binomial distributions with appropriate numbers of trials m_i.

The log-likelihood function of the model is

$$\sum_{i=1}^{14} \left(y_i \ln \left(\frac{\tan^{-1}(\beta_0 + \beta_1 x_i)}{\pi} + \frac{1}{2} \right) + \right.$$
$$\left. (m_i - y_i) \ln \left(\frac{1}{2} - \frac{\tan^{-1}(\beta_0 + \beta_1 x_i)}{\pi} \right) \right),$$

where x_i denotes the value of the index variable for the ith covariate class. ☐

6.3.2 Parametrized link functions

Sometimes a parametrized link function is needed. In the following example the link function is the cumulative distribution of the standard Cauchy distribution raised to an unknown power γ.

Example 6.3 Remission of cancer data (cont'd)

Taking the γth power of the inverse of the cumulative distribution function of the standard Cauchy distribution as the link function means that the inverse link function has the form

$$h(\eta; \gamma) = g^{-1}(\eta; \gamma) = \left(\frac{\tan^{-1}(\eta)}{\pi} + \frac{1}{2} \right)^{\gamma}.$$

Figure 6.5 shows the inverse link function for selected values of γ.

The binomial regression model with this function as the inverse link function is denoted by the expression

$$\mathcal{M} = \text{RegressionModel}\left[X, \left(\frac{\tan^{-1}(\beta_0 x_0 + \beta_1 x_1)}{\pi} + \frac{1}{2} \right)^{\gamma}, \right.$$
$$\{x_0, x_1\}, \{\beta_0, \beta_1, \gamma\},$$
$$\left. \text{Distribution} \rightarrow \{\text{BinomialModel}[2, \omega], \ldots, \text{BinomialModel}[3, \omega]\} \right],$$

where the second argument gives the success probability as the function of the vector (x_0, x_1) of regressors and the vector of parameters $(\beta_0, \beta_1, \gamma)$.

The log-likelihood function of the model is

$$\sum_{i=1}^{14} \left(y_i \gamma \ln \left(\frac{\tan^{-1}(\beta_0 + \beta_1 x_i)}{\pi} + \frac{1}{2} \right) + \right.$$
$$\left. (m_i - y_i) \ln \left(1 - \left(\frac{1}{2} + \frac{\tan^{-1}(\beta_0 + \beta_1 x_i)}{\pi} \right)^{\gamma} \right) \right),$$

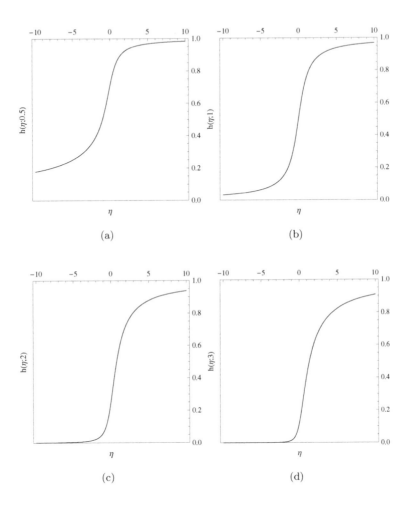

FIGURE 6.5:   Parametrized binomial link function for selected values of γ: (a) 0.5, (b) 1, (c) 2, and (d) 4.

where x_i denotes the value of the index variable for the ith covariate class. $\square$

6.4 Likelihood inference

6.4.1 Likelihood function of binomial data

When responses $y_1, \ldots, y_n$ are assumed to be statistically independent observations from binomial distributions with numbers of trials m_i and success probabilities ω_i, the log-likelihood function has the form

$$\ell(\omega; y) = \sum_{i=1}^{n} \left(y_i \ln \left(\frac{\omega_i}{1 - \omega_i} \right) + m_i \ln (1 - \omega_i) \right), \tag{6.16}$$

where $\omega = (\omega_1, \omega_2, \ldots, \omega_n)$ and $y = (y_1, y_2, \ldots, y_n)$. The relation between factors and vector ω

$$g(\omega_i) = \eta_i = \sum_{j=1}^{d} x_{ij} \beta_j \ \forall i = 1, \ldots, n$$

results into a formula containing parameters $\beta_1, \beta_2, \ldots, \beta_d$.

In the case of the logistic model

$$g(\omega_i) = \eta_i = \ln \left(\frac{\omega_i}{1 - \omega_i} \right) = \sum_{j=1}^{d} x_{ij} \beta_j \ \forall i = 1, \ldots, n.$$

The log-likelihood function now has the form

$$\ell(\beta; y) = \sum_{i=1}^{n} \left[y_i \sum_{j=1}^{d} x_{ij} \beta_j - m_i \ln \left(1 + e^{\sum_{j=1}^{d} x_{ij} \beta_j} \right) \right]$$

$$= \sum_{i=1}^{n} \sum_{j=1}^{d} y_i x_{ij} \beta_j - \sum_{i=1}^{n} m_i \ln \left(1 + e^{\sum_{j=1}^{d} x_{ij} \beta_j} \right),$$

which can be written in the form

$$\ell(\beta; y) = \sum_{j=1}^{d} \beta_j \left(\sum_{i=1}^{n} y_i x_{ij} \right) - \sum_{i=1}^{n} m_i \ln \left(1 + e^{\sum_{j=1}^{d} x_{ij} \beta_j} \right).$$

Thus, the statistics

$$\sum_{i=1}^{n} y_i x_{ij} \ \forall j = 1, \ldots, d$$

together are a minimal sufficient statistic for the parameters β_1, β_2, ..., and β_d, that is,

$$y^\mathrm{T} X$$

is the minimal sufficient statistic.

Example 6.4 O-ring data (cont'd)

Assume that temperature has a linear effect on logit of thermal distress probability, that is, assume that model matrix is the following:

$$X = \begin{pmatrix} 1 & 66 \\ 1 & 70 \\ 1 & 69 \\ \vdots & \vdots \\ 1 & 75 \\ 1 & 76 \\ 1 & 58 \end{pmatrix}.$$

The logistic regression model for frequencies of thermal distress is denoted by

$$\mathcal{M} = \mathrm{LogisticRegressionModel}[X, 6],$$

where the latter argument gives the number of trials which is the same for all components of the response vector. The value of the minimal sufficient statistic is

$$y^\mathrm{T} X = (0, 1, 0, \ldots, 2, 0, 1) \begin{pmatrix} 1 & 66 \\ 1 & 70 \\ 1 & 69 \\ \vdots & \vdots \\ 1 & 79 \\ 1 & 75 \\ 1 & 76 \\ 1 & 58 \end{pmatrix} = (9, 574).$$

▯

6.4.2   Estimates of parameters

Because

$$\frac{\partial \ell}{\partial \omega_i} = \frac{y_i - m_i \omega_i}{\omega_i \left(1 - \omega_i\right)},$$

using the chain rule gives

$$\frac{\partial \ell}{\partial \beta_r} = \sum_{i=1}^{n} \frac{y_i - m_i \omega_i}{\omega_i \left(1 - \omega_i\right)} \frac{\partial \omega_i}{\partial \beta_r}.$$

Using the chain rule again we have

$$\frac{\partial w_i}{\partial \beta_r} = \frac{\partial w_i}{\partial \eta_i}\frac{\partial \eta_i}{\partial \beta_r} = \frac{\partial w_i}{\partial \eta_i}x_{ir},$$

and so

$$\frac{\partial \ell}{\partial \beta_r} = \sum_{i=1}^{n}\frac{y_i - m_i w_i}{w_i\,(1 - w_i)}\frac{\partial w_i}{\partial \eta_i}x_{ir}.$$

Fisher's expected information is

$$-E\left(\frac{\partial^2 \ell}{\partial \beta_r \partial \beta_s}\right) = E\left(\frac{\partial \ell}{\partial \beta_r}\frac{\partial \ell}{\partial \beta_s}\right)$$

$$= \sum_{i=1}^{n}\frac{m_i}{w_i\,(1 - w_i)}\frac{\partial w_i}{\partial \beta_r}\frac{\partial w_i}{\partial \beta_s}$$

$$= \sum_{i=1}^{n}m_i\frac{\left(\frac{\partial w_i}{\partial \eta_i}\right)^2}{w_i\,(1 - w_i)}x_{ir}x_{is}$$

$$= \left\{X^{\mathrm T}WX\right\}_{rs},$$

where W is the diagonal matrix of weights

$$W = \operatorname{diag}\left\{\frac{m_i\left(\frac{\partial w_i}{\partial \eta_i}\right)^2}{w_i\,(1 - w_i)}\right\}.$$

Using this weight matrix in Equation (2.16) produces MLE.
   In the case of logistic regression we get

$$\frac{\partial \ell}{\partial \beta} = X^{\mathrm T}(y - \mu).$$

In addition, the diagonal matrix of weights has the form

$$W = \operatorname{diag}\left\{m_i w_i\,(1 - w_i)\right\}.$$

Example 6.5 O-ring data (cont'd)

   MLEs take the form

$$\begin{pmatrix} 1 & 66 \\ 1 & 70 \\ 1 & 69 \\ \vdots & \vdots \\ 1 & 75 \\ 1 & 76 \\ 1 & 58 \end{pmatrix}^{\mathrm T}\left(\begin{pmatrix} 0 \\ 1 \\ 0 \\ \vdots \\ 2 \\ 0 \\ 1 \end{pmatrix} - \begin{pmatrix} \frac{6e^{\beta_1 + 66\beta_2}}{1 + e^{\beta_1 + 66\beta_2}} \\ \frac{6e^{\beta_1 + 70\beta_2}}{1 + e^{\beta_1 + 70\beta_2}} \\ \frac{6e^{\beta_1 + 69\beta_2}}{1 + e^{\beta_1 + 69\beta_2}} \\ \vdots \\ \frac{6e^{\beta_1 + 75\beta_2}}{1 + e^{\beta_1 + 75\beta_2}} \\ \frac{6e^{\beta_1 + 76\beta_2}}{1 + e^{\beta_1 + 76\beta_2}} \\ \frac{6e^{\beta_1 + 58\beta_2}}{1 + e^{\beta_1 + 58\beta_2}} \end{pmatrix}\right) = \begin{pmatrix} 0 \\ 0 \end{pmatrix}.$$

The solution of the maximum likelihood equations is (5.085, -0.116).   ⬚

6.4.3   Likelihood ratio statistic or deviance function

The log-likelihood of the fitted model is

$$\ell(\hat{\omega}; y) = \sum_{i=1}^{n} \left[y_i \ln(\hat{\omega}_i) + (m_i - y_i) \ln(1 - \hat{\omega}_i) \right].$$

Success probabilities $\tilde{\omega}_i = y_i/m_i$ produce the highest likelihood and thus the deviance function is

$$D(y; \hat{\omega}) = 2\ell(\tilde{\omega}; y) - 2\ell(\hat{\omega}; y)$$

$$= 2 \sum_{i=1}^{n} \left[y_i \ln\left(\frac{y_i}{\hat{\mu}_i}\right) + (m_i - y_i) \ln\left(\frac{m_i - y_i}{m_i - \hat{\mu}_i}\right) \right]. \qquad (6.17)$$

The asymptotic distribution of the deviance $D(y; \hat{\omega})$ is χ^2-distribution with $n - d$ degrees of freedom under the following assumptions:

- Assumption 1: The observations have binomial distributions and are statistically independent.

- Assumption 2: n is fixed and $m_i \to \infty$ for every i, in fact $m_i \omega_i (1 - \omega_i) \to \infty$ for every i.

In the limit $D(y; \hat{\omega})$ is approximately independent of the estimated parameters $\hat{\beta}$ and thus approximately independent of the fitted probabilities $\hat{\omega}$. This independence is essential if D is used as a goodness of fit test statistic.

Deviance function is more useful for comparing nested models than as a measure of goodness of fit.

Example 6.6 O-ring data (cont'd)

Consider the logistic regression model with linear effect of temperature. Log-likelihoods of the full model and logistic regression model are -7.261 and -15.823, respectively. Thus, deviance has the value 17.123 with $23 - 2 = 21$ degrees of freedom. In this example, however, numbers of trials and observed frequencies are so small that according to Assumption 2 the distribution of the deviance cannot be approximated by χ^2-distribution with 21 degrees of freedom.   ⬚

6.4.4   Distribution of deviance

To study by simulation the distribution of deviance consider 10 independent Bernoulli observations with different success probabilities. The model matrix

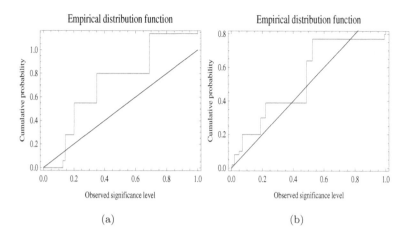

FIGURE 6.6: Empirical distribution functions of the likelihood ratio statistic with alternative model: (a) full and (b) restricted.

of the logistic regression model for the observations is

$$X = \begin{pmatrix} 1\,0\,0\,0\,0\,0\,0\,0\,0\,0 \\ 1\,1\,0\,0\,0\,0\,0\,0\,0\,0 \\ 1\,0\,1\,0\,0\,0\,0\,0\,0\,0 \\ 1\,0\,0\,1\,0\,0\,0\,0\,0\,0 \\ 1\,0\,0\,0\,1\,0\,0\,0\,0\,0 \\ 1\,0\,0\,0\,0\,1\,0\,0\,0\,0 \\ 1\,0\,0\,0\,0\,0\,1\,0\,0\,0 \\ 1\,0\,0\,0\,0\,0\,0\,1\,0\,0 \\ 1\,0\,0\,0\,0\,0\,0\,0\,1\,0 \\ 1\,0\,0\,0\,0\,0\,0\,0\,0\,1 \end{pmatrix}$$

and the logistic regression model is defined by the following expression:

$$\mathcal{M} = \text{LogisticRegressionModel}[X, 1].$$

Consider the statistical hypothesis that all success probabilities are equal. The observed significance level calculated for the response vector consisting of 10 successes is 1.

Next simulate the distribution of observed significance level assuming that the common success probability is $\frac{e}{1+e} = 0.731059$. Figure 6.6 gives the sample distribution function of the observed significance levels. Because the empirical distribution function should be approximately uniform when the distribution of the deviance is well approximated by the appropriate χ^2-distribution, there is a clear indication that in this case the approximation is not good. The calculated observed significance levels tend to be too small.

Next consider 10 independent Bernoulli observations such that the first five have a common success probability and the last five have a different common success probability. The model matrix of the logistic regression model for the observations is

$$X = \begin{pmatrix} 1 & 0 \\ 1 & 0 \\ 1 & 0 \\ 1 & 0 \\ 1 & 0 \\ 1 & 1 \\ 1 & 1 \\ 1 & 1 \\ 1 & 1 \\ 1 & 1 \end{pmatrix}$$

and the logistic regression model is defined by the following expression:

$$\mathcal{M} = \text{LogisticRegressionModel}[X, 1].$$

Consider again the statistical hypothesis that all success probabilities are equal. The observed significance level calculated for the response vector consisting of 10 successes is again 1.

Next simulate again the distribution of observed significance level assuming that the common success probability is $\frac{e}{1+e} = 0.731059$. Figure 6.6 gives the sample distribution function of the observed significance levels. Because the empirical distribution function should be approximately uniform when the distribution of the deviance is well approximated by the appropriate χ^2-distribution, there is a clear indication that in this case the approximation is rather good.

6.4.5 Model checking

Assume that the success probability of the ith component of the response vector is some nonlinear function of the regression parameters

$$\omega_i = \eta_i = \eta_i(\beta).$$

Then the deviance residual of the ith component of the response vector is

$$d_i = \text{sign}(y_i - \hat{\mu}_i) \left[2 \left\{ y_i \ln \left(\frac{y_i}{\hat{\mu}_i} \right) + (m_i - y_i) \ln \left(\frac{m_i - y_i}{m_i - \hat{\mu}_i} \right) \right\} \right]^{1/2}, \quad (6.18)$$

where

$$\hat{\omega}_i = \eta_i(\hat{\beta})$$

and

$$\hat{\mu}_i = m_i \eta_i(\hat{\beta}).$$

The hat matrix is equal to

$$H(\hat{\beta}) = W(\hat{\beta})^{1/2} X(\hat{\beta}) (X(\hat{\beta})^{\mathrm{T}} W(\hat{\beta}) X(\hat{\beta}))^{-1} X(\hat{\beta})^{\mathrm{T}} W(\hat{\beta})^{1/2},$$

where

$$X(\beta) = \frac{\eta(\beta)}{\partial \beta^{\mathrm{T}}} = \left(\frac{\eta_i(\beta)}{\partial \beta_j} \right)$$

and

$$W(\beta) = \mathrm{diag}\left(\frac{m_i}{\eta_i(\beta)(1 - \eta_i(\beta))} \right). \tag{6.19}$$

The standardized deviance residuals have the form

$$r_{Di} = \frac{d_i}{\sqrt{1 - h_{ii}}}. \tag{6.20}$$

The standardized Pearson residual is

$$r_{Pi} = \frac{u_i(\hat{\beta})}{\sqrt{w_i(\hat{\beta})(1 - h_{ii})}} = \frac{y_i - \mu_i}{\sqrt{\mu_i(1 - \mu_i/m_i)(1 - h_{ii})}}. \tag{6.21}$$

Detailed calculations show that the distributions of the quantities

$$r_i^* = r_{Di} + \frac{1}{r_{Di}} \ln\left(\frac{r_{Pi}}{r_{Di}} \right) \tag{6.22}$$

are close to normal.

Example 6.7 O-ring data (cont'd)

Consider the logistic regression model with linear effect of temperature. Figure 6.7 gives QQ-plots of standardized deviance, standardized Pearson, and modified residuals and Figure 6.8 shows scatter plots of modified residuals against temperature, pressure, and fitted values. These plots indicate that the model does not fit very well.   ⬚

6.5   Logistic regression model

The logistic regression model is a binomial regression model such that the mean of the observation y_i can be written in the form

$$\mu_i = m_i \frac{e^{x_{i1}\beta_1 + x_{i2}\beta_2 + \cdots + x_{id}\beta_d}}{1 + e^{x_{i1}\beta_1 + x_{i2}\beta_2 + \cdots + x_{id}\beta_d}} \quad \forall i = 1, \ldots, n, \tag{6.23}$$

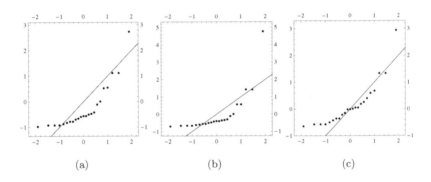

FIGURE 6.7:   O-ring data: QQ-plots of (a) standardized deviance residuals, (b) standardized Pearson residuals, and (c) modified residuals.

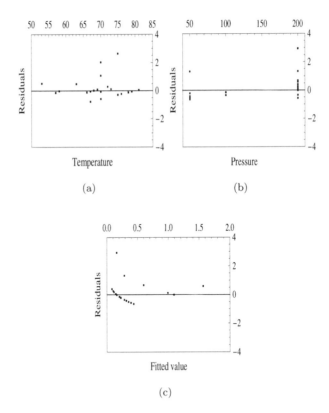

FIGURE 6.8:   O-ring data: Scatter plots of modified residuals against (a) temperature, (b) pressure, and (c) fitted values.

TABLE 6.1: O-ring data: Estimates and approximate 0.95-level confidence intervals of model parameters.

Parameter	Estimate	Lower limit	Upper limit
β_1	5.085	-1.010	11.185
β_2	-0.116	-0.212	-0.0245

where m_i is the number of trials in the binomial distribution of y_i. This means that logit of the success probability of the observation y_i is

$$\ln\left(\frac{\omega_i}{1-\omega_i}\right) = \ln\left(\frac{\mu_i}{m_i - \mu_i}\right) = \eta_i = x_{i1}\beta_1 + x_{i2}\beta_2 + \cdots + x_{id}\beta_d \ \forall i = 1,\ldots,n.$$

Thus, the logistic regression model is a generalized linear model with logit function as link function.

Example 6.8 O-ring data (cont'd)

Consider the logistic regression model with linear effect of temperature. Even though in the previous section plots of residuals indicated that this model is not the perfect model for the data, in the following the statistical inference will be illustrated using this model. Table 6.1 gives MLEs and approximate 0.95-level profile likelihood-based confidence intervals for parameters.

Let us consider the probability of thermal distress at temperature 50, that is, parameter function

$$\psi_1 = \frac{e^{\beta_1 + 50\beta_2}}{1 + e^{\beta_1 + 50\beta_2}}.$$

Its MLE is 0.333 and approximate 0.95-level profile likelihood-based confidence interval is $(0.088, 0.676)$. This interval is considerably large indicating that there is not much information on this interest function. Next consider the difference between this probability and that at temperature 80, that is, parameter function

$$\psi_2 = \frac{e^{\beta_1 + 50\beta_2}}{1 + e^{\beta_1 + 50\beta_2}} - \frac{e^{\beta_1 + 80\beta_2}}{1 + e^{\beta_1 + 80\beta_2}}.$$

Its MLE is 0.318 and approximate 0.95-level profile likelihood-based confidence interval is $(0.047, 0.671)$. Also this interval is considerably large indicating that there is not much information on this interest function, but anyway there is statistical evidence that the thermal distress probability is higher at 50 than that at 80. Figure 6.9 gives plots of log-likelihood functions of all considered interest functions. ☐

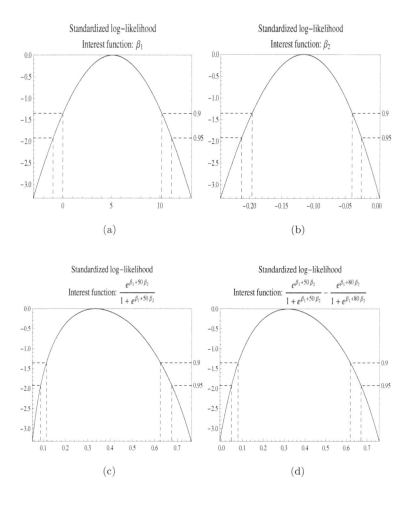

FIGURE 6.9:   O-ring data: Plots of log-likelihoods of interest functions (a) β_1, (b) β_2, (c) ψ_1, and (d) ψ_2.

TABLE 6.2:   Insulin data

Preparation	Dose	Convulsions	Total
Standard	3.4	0	33
Standard	5.2	5	32
Standard	7.0	11	38
Standard	8.5	14	37
Standard	10.5	18	40
Standard	13.0	21	37
Standard	18.0	23	31
Standard	21.0	30	37
Standard	28.0	27	31
Test	6.5	2	40
Test	10.0	10	30
Test	14.0	18	40
Test	21.5	21	35
Test	29.0	27	37

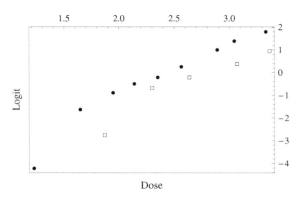

FIGURE 6.10:   Insulin data: Observed logit against dose for standard ($\bullet$) and test ($\square$) preparation.

Example 6.9 **Insulin data**

The effect of insulin on mice, as measured by whether or not they had convulsions, was studied (Lindsey 1995a, p. 68). Columns in Table 6.2 contain values of preparation, frequency of convulsions, and total frequency. Figure 6.10 contains the scatter plot showing dependence of observed logits on dose and preparation. There is a clear effect of dose and that effect seems to be the same for both preparations with the standard preparation having higher log-odds than the test preparation.

TABLE 6.3: Insulin data: Estimates and approximate 0.95-level confidence intervals of model parameters.

Parameter	Estimate	Lower limit	Upper limit
β_1	-5.474	-6.568	-4.457
β_2	-0.899	-1.361	-0.452
β_3	2.257	1.847	2.699

The logistic regression model for frequency of convulsions with linear effect of log-dose and preparation is defined by the model matrix

$$X = \begin{pmatrix} 1\ 0\ 1.224 \\ 1\ 0\ 1.649 \\ 1\ 0\ 1.946 \\ \vdots\ \vdots\ \ \ \vdots \\ 1\ 1\ 2.639 \\ 1\ 1\ 3.068 \\ 1\ 1\ 3.367 \end{pmatrix}$$

and the expression

$$\mathcal{M} = \text{LogisticRegressionModel}[X, \{33, 32, 38, \ldots, 40, 35, 37\}].$$

MLEs and approximate 0.95-level profile likelihood-based confidence intervals for parameters are given in Table 6.3. Both preparation and dose have clear effects.

Figure 6.11 gives plots of residuals which indicate slight deviance from the model. The reason for this is the small curvature in the observed logits.

The probability of convulsions for the standard preparation at dose x has the form

$$\omega(x; \beta_1, \beta_2) = \frac{e^{\beta_1 + \beta_3 x}}{1 + e^{\beta_1 + \beta_3 x}}.$$

Consider the so-called LD50 of the standard preparation, that is, the interest function

$$\psi_1 = -\frac{\beta_1}{\beta_3}.$$

The MLE and approximate 0.95-level profile likelihood-based confidence interval are 2.425 and $(2.308, 2.544)$, respectively. In addition, consider the difference between LD50s of groups, that is, the interest function

$$\psi_2 = \frac{\beta_1}{\beta_3} - \frac{\beta_1 + \beta_2}{\beta_3}.$$

The MLE and approximate 0.95-level profile likelihood-based confidence interval are 0.398 and $(0.207, 0.593)$, respectively.

Figure 6.12 gives plots of log-likelihood functions of all considered interest functions. ☐

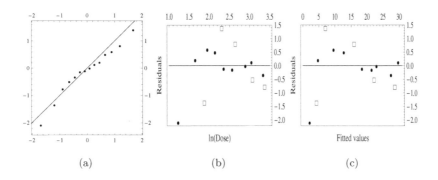

FIGURE 6.11: Insulin data: (a) QQ-plot of modified residuals, (b) scatter plot of modified residuals against logarithm of dose, and (c) scatter plot of modified residuals against fitted values.

6.6 Models with other link functions

Binomial regression model with link function g and inverse link function $h = g^{-1}$ is such that the mean of the observation y_i can be written in the form

$$\mu_i = m_i h \left(x_{i1}\beta_1 + x_{i2}\beta_2 + \cdots + x_{id}\beta_d \right) \ \forall i = 1, \ldots, n,$$

where m_i is the number of trials in the binomial distribution of y_i. This means that for the success probability of the observation y_i it is true that

$$g\left(\omega_i\right) = \eta_i = x_{i1}\beta_1 + x_{i2}\beta_2 + \cdots + x_{id}\beta_d \ \forall i = 1, \ldots, n.$$

***Example 6.10* Remission of cancer data (cont'd)**

Consider a binomial regression model with probit link and with dose having linear effect, that is, model matrix

$$X = \begin{pmatrix} 1 & 8 \\ 1 & 10 \\ 1 & 12 \\ \vdots & \vdots \\ 1 & 32 \\ 1 & 34 \\ 1 & 38 \end{pmatrix}$$

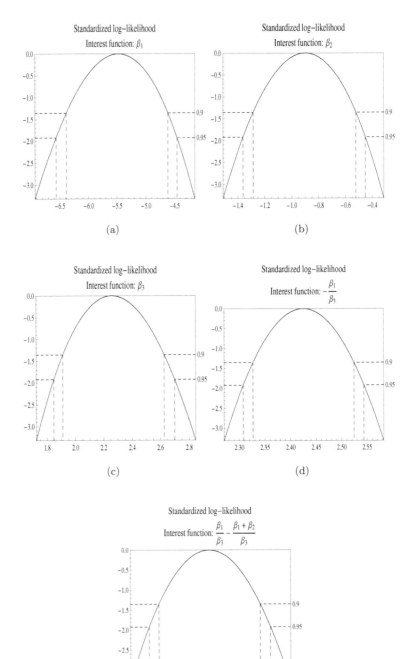

FIGURE 6.12: Insulin data: Plots of log-likelihoods of interest functions (a) β_1, (b) β_2, (c) β_3, (d) ψ_1, and (e) ψ_2.

TABLE 6.4: Remission of cancer data: Estimates and approximate 0.95-level confidence intervals of model parameters.

Parameter	Estimate	Lower limit	Upper limit
β_0	-2.319	-4.011	-0.908
β_1	0.0879	0.0275	0.157

and statistical model defined by

$$\mathcal{M} = \text{GeneralizedLinearRegressionModel}[X,$$
$$\text{Distribution} \rightarrow \{\text{BinomialModel}[2, \gamma], \ldots, \text{BinomialModel}[3, \gamma]\},$$
$$\text{LinkFunction} \rightarrow \text{Probit}].$$

MLEs and approximate 95%-level profile likelihood-based confidence intervals for parameters are given in Table 6.4.

Figure 6.13 gives plots of residuals which indicate a slight deviance from the model. The reason for this is the small curvature in the observed logits. □

6.7   Nonlinear binomial regression model

The nonlinear binomial regression model is a binomial regression model such that the mean of the observation y_i can be written in the form

$$\mu_i = m_i \eta_i(\beta_1, \beta_2, \cdots, \beta_d) \ \forall i = 1, \ldots, n,$$

where m_i is the number of trials in the binomial distribution of y_i and η_i is some nonlinear function of regression parameters β_1, β_2, ..., and β_d.

Example 6.11 **Malaria data**

Proportion of individuals in Bongono, Zaire, with malaria antibodies present was recorded (Morgan, 1992, p. 16; Lindsey 1995a, p. 83). Columns in Table 6.5 contain values of mean age, frequency of seropositive, and total frequency. Figure 6.14 gives a plot of logits against mean age. There is a clear effect of mean age on log-odds. The relation between log-odds and mean age, however, might be nonlinear.

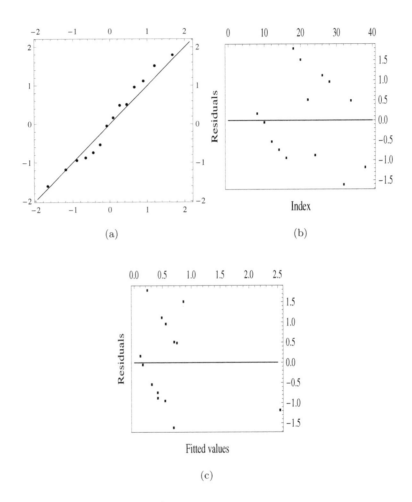

FIGURE 6.13: Remission of cancer data: (a) QQ-plot of modified residuals, (b) scatter plot of modified residuals against index, and (c) scatter plot of modified residuals against fitted values.

TABLE 6.5: Malaria data

Mean age	Seropositive	Total
1.0	2	60
2.0	3	63
3.0	3	53
4.0	3	48
5.0	1	31
7.3	18	182
11.9	14	140
17.1	20	138
22.0	20	84
27.5	19	77
32.0	19	58
36.8	24	75
41.6	7	30
49.7	25	62
60.8	44	74

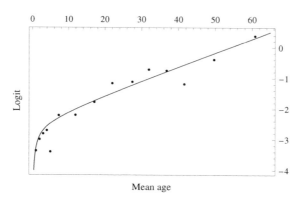

FIGURE 6.14: Malaria data: Scatter plot of observations including the fitted regression function.

Consider a nonlinear binomial regression model with model matrix

$$X = \begin{pmatrix} 1.0 \\ 2.0 \\ 3.0 \\ \vdots \\ 41.6 \\ 49.7 \\ 60.8 \end{pmatrix}$$

and regression function

$$\eta(x_i; \beta_1, \beta_2, \beta_3) = \frac{e^{\beta_1 + x_i \beta_2 - \frac{1}{x_i^{\beta_3}}}}{1 + e^{\beta_1 + x_i \beta_2 - \frac{1}{x_i^{\beta_3}}}} \quad \forall i = 1, \ldots, n.$$

The statistical model is defined by the following expression:

$$\mathcal{M} = \text{RegressionModel}\left[X, \eta(x; \beta_1, \beta_2, \beta_3), \{x\}, \{\beta_1, \beta_2, \beta_3\},\right.$$
$$\left. \text{Distribution} \to \{\text{BinomialModel}[60, \omega], \ldots, \text{BinomialModel}[74, \omega]\}\right].$$

The MLE is (-2.318, 0.0446, 0.603). Approximate 0.95-level profile likelihood-based confidence interval for β_2 is (0.0357, 0.0557).

$\square$

6.8　Bibliographic notes

Important books on binomial regression are Cox (1970), Cox and Snell (1989), Collet (1991), and Morgan (1992).

Chapter 7

Poisson Regression Model

7.1 Data

Suppose that a response y_i has been observed for N statistical units. A basic assumption in this chapter is the following:

$$\text{var}(y_i) = \sigma^2 \text{E}(y_i) \ \forall i = 1, \ldots, N. \tag{7.1}$$

If the possible values of responses consist of nonnegative integers and if the responses are observations from Poisson distributions, a special case with $\sigma^2 = 1$ arises.

For the statistical unit i let

$$\text{E}(y_i) = \mu_i \ \forall i = 1, \ldots, N.$$

Usually the mean for unit i is supposed to depend on values $(x_{i1}, \ldots, x_{id})$ of factors connected with the unit, that is,

$$\text{E}(y_i) = \mu_i = \mu(x_{i1}, \ldots, x_{id}) \ \forall i = 1, \ldots, N,$$

where $\mu(x_1, \ldots, x_d)$ is a given function of the factors.

Assume now that the function $\mu(x_1, \ldots, x_q)$ in addition to factors depends on d unknown common parameters $\beta_1, \ldots, \beta_d$, that is,

$$\mu_i(\beta_1, \ldots, \beta_d) = \mu(x_{i1}, \ldots, x_{iq}; \beta_1, \ldots, \beta_d) \ \forall i = 1, \ldots, n. \tag{7.2}$$

Then the model is a GRM. We call it a *Poisson regression model* in general. This includes cases where means depend nonlinearly on regression parameters. If the function μ has the form

$$\mu(x_{i1}, \ldots, x_{id}; \beta_1, \ldots, \beta_d) = h(x_{i1}\beta_1 + x_{i2}\beta_2 + \cdots + x_{id}\beta_d) \ \forall i = 1, \ldots, n \tag{7.3}$$

for some known function h, the model is a GLIM. The function h is the inverse of the link function, that is, $g = h^{-1}$. Finally, when $h(z) = e^z$ the model is called a *log-linear model*.

Example 7.1 Lizard data

Table A.13 on page 260 contains a comparison of site preferences of two species of lizard, *Grahami* and *Opalinus* (Schoener 1970; McCullagh and

Nelder 1989, p. 128). The columns of the table contain values of illumination, perch diameter in inches, perch height in inches, time of day, species, and number of lizards. Counts in the table may be modeled as independent Poisson observations with means depending on five explaining variables.

Figure 7.1 gives plots showing the dependence of logits on levels of factors. Figure 7.1 (a) shows logarithms of counts so that the first 12 counts are for the case of illumination *Sun*, the first 6 and from 13 to 18 are for the case of perch diameter "≤ 2," the first 3, from 7 to 9, from 13 to 15, and from 19 to 21 are for case of perch height "< 5," and finally every third starting from one are for the case of time of day *Early*, etc. From this plot it is seen that lizards of species *Grahami* are altogether somewhat more numerous than lizards of species *Opalinus* and both species like shade more than sun. In addition from plots indicating "main" effects of factors one can see that time of day seems to have an effect such that at time *Midday* lizards are more numerous than at other times, although this seems to apply only for lizards in shade, and more lizards seem to appear at time *Early* for lizards in the sun.   ⬚

7.2   Poisson distribution

The moment generating function of PoissonModel[μ] has the form

$$M(\xi) = E(e^{\xi y}) = e^{(e^{\xi}-1)\mu} \tag{7.4}$$

and its cumulant generating function is

$$K(\xi) = \ln(M(\xi)) = (e^{\xi} - 1)\mu. \tag{7.5}$$

The first six moments are

$$m_1 = \mu \tag{7.6}$$
$$m_2 = \mu(\mu + 1) \tag{7.7}$$
$$m_3 = \mu(\mu^2 + 3\mu + 1) \tag{7.8}$$
$$m_4 = \mu(\mu^3 + 6\mu^2 + 7\mu + 1) \tag{7.9}$$
$$m_5 = \mu(\mu^4 + 10\mu^3 + 25\mu^2 + 15\mu + 1) \tag{7.10}$$
$$m_6 = \mu(\mu^5 + 15\mu^4 + 65\mu^3 + 90\mu^2 + 31\mu + 1 \tag{7.11}$$

The first four cumulants of the PoissonModel[μ] are

$$\kappa_1 = \mu, \tag{7.12}$$
$$\kappa_2 = \mu, \tag{7.13}$$
$$\kappa_3 = \mu, \tag{7.14}$$
$$\kappa_4 = \mu. \tag{7.15}$$

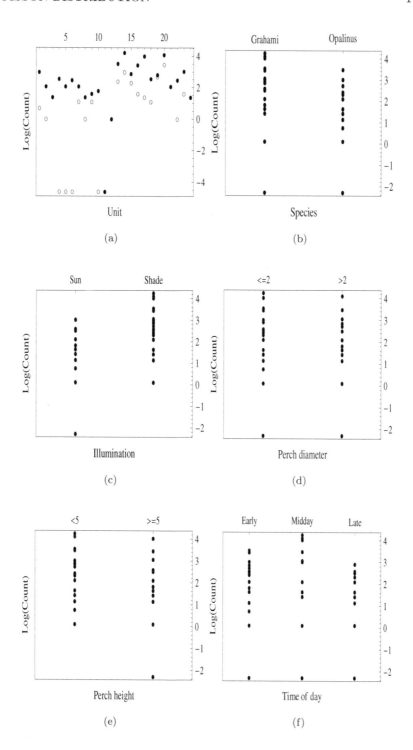

FIGURE 7.1:   Lizard data: Scatter plot of logarithms of counts of lizards against factors. (a) all cells, (b) species, (c) illumination, (d) perch diameter, (e) perch height, and (e) time of day.

In actual fact all cumulants of the PoissonModel[μ] are equal to the mean μ, which can be seen from the cumulant generating function.

The cumulants of standardized variable

$$z = \frac{y - \mu}{\sqrt{\mu}}$$

are seen from the Taylor series expansion of the cumulant generating function

$$K_z(\xi) = \frac{\xi^2}{2} + \frac{\xi^3}{6\sqrt{\mu}} + \frac{\xi^4}{24\mu} + \frac{\xi^5}{120\mu^{3/2}} + \frac{\xi^6}{720\mu^2} + O\left(\xi^7\right)$$

to be of order 0, 1, $O\left(\mu^{-1/2}\right)$, $O\left(\mu^{-1}\right)$, etc. When $r \geq 2$, r^{th} cumulant of z is of order $O\left(\mu^{1-r/2}\right)$. The cumulants of z approach the cumulants 0, 1, 0, 0, ... of standardized normal distribution as $\mu \to \infty$. Because convergence of cumulants implies convergence in distribution, the approximate probabilities will be obtained from the formulae

$$\Pr(y \geq \tilde{y}) \approx 1 - \Phi(z^-)$$

and

$$\Pr(Y \leq \tilde{y}) \approx \Phi(z^+),$$

where Φ is the cumulative distribution function of the standard normal distribution, $\tilde{y}$ is an integer and

$$z^- = \frac{\tilde{y} - \mu - \frac{1}{2}}{\sqrt{\mu}} \quad \text{and} \quad z^+ = \frac{\tilde{y} - \mu + \frac{1}{2}}{\sqrt{\mu}}.$$

Thus, the effect of the continuity correction $\mp\frac{1}{2}$ on the probabilities is of order $O\left(\mu^{-1/2}\right)$ and so asymptotically negligible. In medium size of mean the effect of the continuity correction is remarkable and almost always improves the approximation.

Figure 7.2 contains graphs of point probability functions of Poisson distributions with means 5, 10, 20, 30, and 50.

7.3   Link functions

7.3.1   Unparametrized link functions

The commonly used link function is the log-link, that is,

$$g(\mu) = \ln(\mu). \tag{7.16}$$

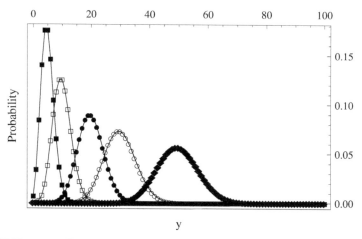

FIGURE 7.2: Point probability function of Poisson distribution with means 5 (■), 10 (□), 20 (•), 30 (∘), and 50 (♦).

Given two distributions with cumulative distribution functions F_1 and F_2 such that the first distribution can have only positive values and the second can have any real number as a value, then the function

$$g(\omega) = F_2^{-1}(F_1(\mu)) \qquad (7.17)$$

is a possible link function.

In case of log-link the distributions are the standardized log-normal and normal distributions because

$$F_2^{-1}(F_1(\mu)) = \Phi^{-1}(\Phi(\ln(\mu))) = \ln(\mu).$$

Thus, log-link might be called log-normal – normal link function.

For example, taking F_1 to be the cumulative distribution function of the exponential distribution with mean one and F_2 the standardized normal distribution, we get

$$F_2^{-1}(F_1(\mu)) = \Phi^{-1}(1 - e^{-\mu}).$$

Figure 7.3 compares the previous two link functions.

In the next example the cumulative distribution function of a Cauchy distribution with location 0 and scale 1 is used together with exponential distribution, that is, the exponential – Cauchy link function is used.

Example 7.2 Stressful events data (cont'd)

Assuming that the linear predictor contains, in addition to a constant, the

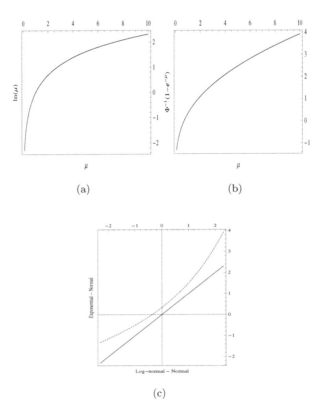

FIGURE 7.3: Comparison of link functions for Poisson distribution. (a) Log-normal – normal, (b) exponential – Cauchy, and (c) comparison of link functions.

month linearly, the design matrix is

$$X = \begin{pmatrix} 1 & 1 \\ 1 & 2 \\ 1 & 3 \\ \vdots & \vdots \\ 1 & 16 \\ 1 & 17 \\ 1 & 18 \end{pmatrix}.$$

The exponential – Cauchy link function is defined by the expression

$$h(z) = -\ln\left(\frac{1}{2} - \frac{\tan^{-1}(z)}{\pi}\right) \forall z \in \mathbb{R}.$$

A Poisson regression model with the exponential – Cauchy link function for the first three counts is defined by

$$\mathcal{M} = \text{RegressionModel}\left[\begin{pmatrix} 1 & 1 \\ 1 & 2 \\ 1 & 3 \end{pmatrix},\right.$$

$$\left.\text{Distribution} \rightarrow \text{PoissonModel}[\mu], \text{InverseLink} \rightarrow h\right].$$

Its log-likelihood function up to a constant has the following form:

$$\ln\left(\frac{1}{2} - \frac{\tan^{-1}(\beta_1 + \beta_2)}{\pi}\right) + \ln\left(\frac{1}{2} - \frac{\tan^{-1}(\beta_1 + 2\beta_2)}{\pi}\right)$$

$$+ \ln\left(\frac{1}{2} - \frac{\tan^{-1}(\beta_1 + 3\beta_2)}{\pi}\right) + 15\ln\left(-\ln\left(\frac{1}{2} - \frac{\tan^{-1}(\beta_1 + \beta_2)}{\pi}\right)\right)$$

$$+ 11\ln\left(-\ln\left(\frac{1}{2} - \frac{\tan^{-1}(\beta_1 + 2\beta_2)}{\pi}\right)\right) +$$

$$14\ln\left(-\ln\left(\frac{1}{2} - \frac{\tan^{-1}(\beta_1 + 3\beta_2)}{\pi}\right)\right).$$

□

7.3.2 Parametrized link functions

Example 7.3 Stressful events data (cont'd)

Consider the following parametrized inverse link function:

$$h(\eta; \gamma) = F_1^{-1}(F_2(\eta)) = -\ln(1 - \Phi(\eta)^\gamma).$$

This defines a Poisson regression model with parametrized link function h for the first three counts

$$\mathcal{M} = \text{RegressionModel}\left[\begin{pmatrix} 1 & 1 \\ 1 & 2 \\ 1 & 3 \end{pmatrix}, -\ln(1 - \Phi(x_1\beta_1 + x_2\beta_2)^\gamma), \{x_1, x_2\}, \right.$$

$$\left. \{\gamma, \beta_1, \beta_2\}, \text{Distribution} \rightarrow \text{PoissonModel}[\mu]\right].$$

The log-likelihood function of the first three counts has the form:

$$\ln\left(1 - \Phi\left(\beta_1 + \beta_2\right)^\gamma\right) + \ln\left(1 - \Phi\left(\beta_1 + 2\beta_2\right)^\gamma\right) + \ln\left(1 - \Phi\left(\beta_1 + 3\beta_2\right)^\gamma\right)$$
$$+ 15\ln\left(-\ln\left(1 - \Phi\left(\beta_1 + \beta_2\right)^\gamma\right)\right) + 11\ln\left(-\ln\left(1 - \Phi\left(\beta_1 + 2\beta_2\right)^\gamma\right)\right)$$
$$+ 14\ln\left(-\ln\left(1 - \Phi\left(\beta_1 + 3\beta_2\right)^\gamma\right)\right).$$

$\square$

7.4 Likelihood inference

7.4.1 Likelihood function of Poisson data

When responses $y_1, \ldots, y_n$ are assumed to be statistically independent observations from Poisson distributions with means μ_i, the log-likelihood function has the form

$$\ell(\omega; y) = \sum_{i=1}^n (y_i \ln(\mu_i) - \mu_i - \ln(y_i!)) \tag{7.18}$$

where $\omega = (\mu_1, \mu_2, \ldots, \mu_n)$ and $y = (y_1, y_2, \ldots, y_n)$. Relation between factors and vector ω

$$g(\mu_i) = \eta_i = \sum_j x_{ij}\beta_j \ \forall i = 1, \ldots, n$$

results in a formula containing parameters $\beta_1, \beta_2, \ldots, \beta_d$.

In the case of a log-linear model

$$g(\mu_i) = \ln(\mu_i) = \eta_i = \sum_j x_{ij}\beta_j \forall i = 1, \ldots, n.$$

The log-likelihood function now has the form

$$\ell(\beta; y) = \sum_i \left[y_i \sum_j x_{ij}\beta_j + e^{\sum_j x_{ij}\beta_j}\right] = \sum_{i,j} y_i x_{ij}\beta_j + \sum_i e^{\sum_j x_{ij}\beta_j},$$

which can be written in the form

$$\ell(\beta; y) = \sum_j \beta_j \sum_i y_i x_{ij} + \sum_i e^{\sum_j x_{ij}\beta_j}.$$

Thus, the statistics

$$\sum_i y_i x_{ij} \forall j = 1, \ldots, d$$

together are a minimal sufficient statistic for the parameters $\beta_1, \beta_2, \ldots, \beta_d$.

Example 7.4 Stressful events data (cont' d)

Assume that month has linear effect on log of mean count of stressful events, that is, assume that the model matrix is the following:

$$X = \begin{pmatrix} 1 & 1 \\ 1 & 2 \\ 1 & 3 \\ \vdots & \vdots \\ 1 & 16 \\ 1 & 17 \\ 1 & 18 \end{pmatrix}.$$

The log-linear regression model for counts of stressful events is denoted by

$$\mathcal{M} = \text{PoissonRegressionModel}[X].$$

The value of the minimal sufficient statistic is

$$y^T X = (15, 11, 14, \ldots, 1, 1, 4) \begin{pmatrix} 1 & 1 \\ 1 & 2 \\ 1 & 3 \\ \vdots & \vdots \\ 1 & 16 \\ 1 & 17 \\ 1 & 18 \end{pmatrix} = (147, 1077).$$

☐

7.4.2 Estimates of parameters

Because

$$\frac{\partial \ell}{\partial \mu_i} = \frac{y_i}{\mu_i} - 1,$$

using chain rule gives

$$\frac{\partial \ell}{\partial \beta_r} = \sum_{i=1}^{n} \left(\frac{y_i}{\mu_i} - 1 \right) \frac{\partial \mu_i}{\partial \beta_r}.$$

Using chain rule again we have

$$\frac{\partial \mu_i}{\partial \beta_r} = \frac{\partial \mu_i}{\partial \eta_i} \frac{\partial \eta_i}{\partial \beta_r} = \frac{\partial \mu_i}{\partial \eta_i} x_{ir},$$

and so

$$\frac{\partial \ell}{\partial \beta_r} = \sum_{i=1}^{n} \left(\frac{y_i}{\mu_i} - 1 \right) \frac{\partial \mu_i}{\partial \eta_i} x_{ir}.$$

Fisher's expected information is

$$
\begin{aligned}
-\mathrm{E} \left(\frac{\partial^2 \ell}{\partial \beta_r \partial \beta_s} \right) &= \sum_{i=1}^{n} \frac{1}{\mu_i} \frac{\partial \mu_i}{\partial \beta_r} \frac{\partial \mu_i}{\partial \beta_s} \\
&= \sum_{i=1}^{n} \frac{\left(\frac{\partial \mu_i}{\partial \eta_i} \right)^2}{\mu_i} x_{ir} x_{is} \\
&= \left\{ X^{\mathrm{T}} W X \right\}_{rs},
\end{aligned}
$$

where W is the diagonal matrix of weights

$$W = \mathrm{diag} \left\{ \frac{\left(\frac{\partial \mu_i}{\partial \eta_i} \right)^2}{\mu_i} \right\}.$$

Using this weight matrix in Equation (2.16) produces MLE.

In the case of log-linear regression we get

$$\frac{\partial \ell}{\partial \beta} = X^{\mathrm{T}} (y - \mu).$$

In addition, the diagonal matrix of weights has the form

$$W = \mathrm{diag} \left\{ \mu_i \right\}.$$

Example 7.5 **Stressful events data (cont' d)**

Maximum likelihood equations thus take the form

$$
\begin{pmatrix} 1 & 1 \\ 1 & 2 \\ 1 & 3 \\ \vdots & \vdots \\ 1 & 16 \\ 1 & 17 \\ 1 & 18 \end{pmatrix}^{T} \left(\begin{pmatrix} 15 \\ 11 \\ 14 \\ \vdots \\ 1 \\ 1 \\ 4 \end{pmatrix} - \begin{pmatrix} e^{\beta_1+\beta_2} \\ e^{\beta_1+2\beta_2} \\ e^{\beta_1+3\beta_2} \\ \vdots \\ e^{\beta_1+16\beta_2} \\ e^{\beta_1+17\beta_2} \\ e^{\beta_1+18\beta_2} \end{pmatrix} \right) = \begin{pmatrix} 0 \\ 0 \end{pmatrix}.
$$

The solution of the maximum likelihood equations is (2.803, -0.0838). ☐

7.4.3   Likelihood ratio statistic or deviance function

In the fitted model the log-likelihood is

$$
\ell(\hat{\omega}; y) = \sum_{i=1}^{n} (y_i \ln(\hat{\mu}_i) - \hat{\mu}_i).
$$

The values $\tilde{\mu}_i = y_i$ produce the highest likelihood. Thus, the deviance function is

$$
D(y; \hat{\omega}) = 2\ell(\tilde{\mu}; y) - 2\ell(\hat{\mu}; y) = 2 \sum_i \left(y_i \ln \left(\frac{y_i}{\hat{\mu}_i} \right) - (y_i - \hat{\mu}_i) \right). \qquad (7.19)
$$

The asymptotic distribution of the deviance $D(y; \hat{\omega})$ is χ^2-distribution with $n - d$ degrees of freedom.

Example 7.6 **Stressful events data (cont'd)**

Consider the log-linear regression model with the linear effect of month. The log-likelihood of the full model and log-linear regression model are -33.6272 and -45.9124, respectively. Thus, deviance has the value of 24.5704 with $18 - 2 = 16$ degrees of freedom. ☐

7.4.4   Distribution of deviance

To study by simulation the distribution of deviance, consider 10 independent Poisson observations with different means. The design matrix of the log-linear regression model for the observations is

$$X = \begin{pmatrix} 1\,0\,0\,0\,0\,0\,0\,0\,0\,0 \\ 1\,1\,0\,0\,0\,0\,0\,0\,0\,0 \\ 1\,0\,1\,0\,0\,0\,0\,0\,0\,0 \\ 1\,0\,0\,1\,0\,0\,0\,0\,0\,0 \\ 1\,0\,0\,0\,1\,0\,0\,0\,0\,0 \\ 1\,0\,0\,0\,0\,1\,0\,0\,0\,0 \\ 1\,0\,0\,0\,0\,0\,1\,0\,0\,0 \\ 1\,0\,0\,0\,0\,0\,0\,1\,0\,0 \\ 1\,0\,0\,0\,0\,0\,0\,0\,1\,0 \\ 1\,0\,0\,0\,0\,0\,0\,0\,0\,1 \end{pmatrix}$$

and Poisson regression model is defined by the following expression:

$$\mathcal{M} = \text{PoissonRegressionModel}[X].$$

Observed significance level for the statistical hypothesis that all means are equal calculated for the response vector consisting of counts 0, 1, 2, 3, 4, 5, 6, 7, 8, and 9 is 0.00679198.

Next simulate the distribution of the observed significance level assuming that the common mean is $e = 2.71828$. Figure 7.4 gives the sample distribution function of the observed significance levels. Because the empirical distribution function should be approximately uniform when the distribution of the deviance is well approximated by the appropriate χ^2-distribution, there is some indication that in this case the approximation is not that good. The calculated observed significance levels tend to be too small.

Next consider 10 independent Poisson observations such that the first five observations have common mean and last five observations have different common mean. The design matrix of the log-linear regression model for the observations is

$$X = \begin{pmatrix} 1\,0 \\ 1\,0 \\ 1\,0 \\ 1\,0 \\ 1\,0 \\ 1\,1 \\ 1\,1 \\ 1\,1 \\ 1\,1 \\ 1\,1 \end{pmatrix}$$

and the Poisson regression model is defined by the following expression

$$\mathcal{M} = \text{PoissonRegressionModel}[X].$$

The observed significance level for the statistical hypothesis that all means are equal calculated for the response vector consisting of counts 0, 1, 2, 3, 4, 5, 6, 7, 8, and 9 is 0.0001254.

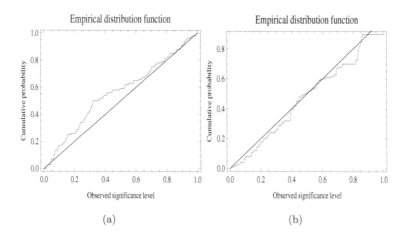

FIGURE 7.4: Empirical distribution functions of the likelihood ratio statistic for (a) full alternative and (b) restricted alternative hypothesis.

Next simulate the distribution of the observed significance level assuming that the common mean is $e = 2.71828$. Figure 7.4 gives the sample distribution function of the observed significance levels. Because the empirical distribution function should be approximately uniform when the distribution of the deviance is well approximated by the appropriate χ^2-distribution, there is clear indication that in this case the approximation is rather good.

7.4.5   Model checking

Assume that the success probability of the ith component of the response vector is some nonlinear function of the regression parameters

$$\mu_i = \eta_i = \eta_i(\beta).$$

Then the deviance residual of the ith component of the response vector is

$$d_i = \text{sign}(y_i - \hat{\mu}_i)\sqrt{2\left(y_i \ln\left(\frac{y_i}{\hat{\mu}_i}\right) - (y_i - \hat{\mu}_i)\right)} \qquad (7.20)$$

where

$$\hat{\mu}_i = \eta_i(\hat{\beta}).$$

The hat matrix is equal to

$$H(\hat{\beta}) = W(\hat{\beta})^{1/2} X(\hat{\beta}) (X(\hat{\beta})^{\mathrm{T}} W(\hat{\beta}) X(\hat{\beta}))^{-1} X(\hat{\beta})^{\mathrm{T}} W(\hat{\beta})^{1/2},$$

where
$$X(\beta) = \frac{\eta(\beta)}{\partial \beta^{\mathrm{T}}} = \left(\frac{\eta_i(\beta)}{\partial \beta_j}\right)$$

and
$$W(\beta) = \mathrm{diag}\left(\frac{1}{\eta_i(\beta)}\right).$$

The standardized deviance residuals have the form

$$r_{Di} = \frac{d_i}{\sqrt{1 - h_{ii}}}, \tag{7.21}$$

The standardized Pearson residual is

$$r_{Pi} = \frac{u_i(\hat{\beta})}{\sqrt{w_i(\hat{\beta})(1 - h_{ii})}} = \frac{y_i - \eta_i}{\sqrt{\eta_i(1 - h_{ii})}}. \tag{7.22}$$

Detailed calculations show that the distributions of the quantities

$$r_i^* = r_{Di} + \frac{1}{r_{Di}} \ln\left(\frac{r_{Pi}}{r_{Di}}\right) \tag{7.23}$$

are close to standard normal.

Example 7.7 Stressful events data (cont'd)

Figure 7.5 gives standardized deviance, Pearson, and modified residuals for
the log-linear model.   □

7.5   Log-linear model

The log-linear model is a Poisson regression model such that the mean of
the observation y_i can be written in the form

$$\mu_i = e^{x_{i1}\beta_1 + x_{i2}\beta_2 + \cdots + x_{id}\beta_d} \ \forall i = 1, \ldots, n. \tag{7.24}$$

This means that the logarithm of the mean of the observation y_i is

$$\ln(\mu_i) = \eta_i = x_{i1}\beta_1 + x_{i2}\beta_2 + \cdots + x_{id}\beta_d \ \forall i = 1, \ldots, n.$$

Thus, the log-linear model is a generalized linear model with the logarithm
function as the link function.

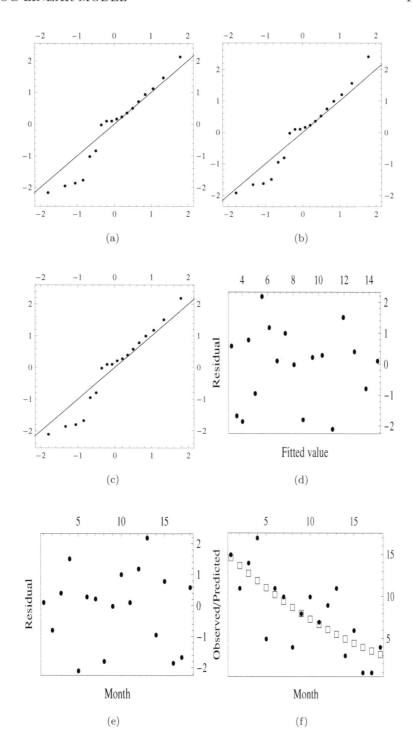

FIGURE 7.5: Stressful events data: (a) deviance residuals, (b) Pearson residuals, (c) modified residuals, (d) residual against fitted values, (e) residuals against month, and (f) observed and predicted values against month.

Example 7.8 Lizard data (cont'd)

The model matrix including all effects is

$$
X = \begin{pmatrix}
1\,1\,1\,1\,1\,1\,0\,1\,1\,1 & \cdots & 1\,0\,1\,0\,1\,0\,1\,0\,1\,0 \\
1\,0\,1\,1\,1\,1\,0\,0\,0\,0 & \cdots & 0\,0\,0\,0\,0\,0\,1\,0\,0\,0 \\
1\,1\,1\,1\,1\,0\,1\,1\,1\,1 & \cdots & 0\,1\,0\,1\,0\,1\,0\,1\,0\,1 \\
\vdots\,\vdots\,\vdots\,\vdots\,\vdots\,\vdots\,\vdots\,\vdots\,\vdots\,\vdots & \ddots & \vdots\,\vdots\,\vdots\,\vdots\,\vdots\,\vdots\,\vdots\,\vdots\,\vdots\,\vdots \\
1\,0\,0\,0\,0\,0\,1\,0\,0\,0 & \cdots & 0\,0\,0\,0\,0\,0\,0\,0\,0\,0 \\
1\,1\,0\,0\,0\,0\,0\,0\,0\,0 & \cdots & 0\,0\,0\,0\,0\,0\,0\,0\,0\,0 \\
1\,0\,0\,0\,0\,0\,0\,0\,0\,0 & \cdots & 0\,0\,0\,0\,0\,0\,0\,0\,0\,0
\end{pmatrix}.
$$

AIC for the full model is 279.0 and those for models with four, three, and two factor interactions are 275.3, 267.0, and 241.3, respectively. Finally AIC for the model containing only main effects is 325.4. Thus, the model containing only two factor interactions is selected for more detailed analysis.

The following model matrix

$$
X_1 = \begin{pmatrix}
1\,1\,1\,1\,1\,1\,0\,1\,1\,1\,1\,0\,1\,1\,1\,0\,1\,1\,0\,1\,0 \\
1\,0\,1\,1\,1\,1\,0\,0\,0\,0\,0\,0\,1\,1\,1\,0\,1\,1\,0\,1\,0 \\
1\,1\,1\,1\,1\,0\,1\,1\,1\,1\,0\,1\,1\,1\,0\,1\,1\,0\,1\,0\,1 \\
\vdots\,\vdots \\
1\,0\,0\,0\,0\,0\,1\,0\,0\,0\,0\,0\,0\,0\,0\,0\,0\,0\,0\,0\,0 \\
1\,1\,0\,0\,0\,0\,0\,0\,0\,0\,0\,0\,0\,0\,0\,0\,0\,0\,0\,0\,0 \\
1\,0\,0\,0\,0\,0\,0\,0\,0\,0\,0\,0\,0\,0\,0\,0\,0\,0\,0\,0\,0
\end{pmatrix}
$$

with

$$\mathcal{M}_1 = \text{PoissonRegressionModel}\,[X_1]$$

defines the log-linear regression model with effects up to two factor interactions.

Table 7.1 gives MLEs and approximate 0.95-level profile likelihood-based confidence intervals of model parameters.

The following model matrix

$$
X_2 = \begin{pmatrix}
1\,1\,1\,1\,1\,1\,0\,1\,1\,1\,1\,0\,1\,0\,1 \\
1\,0\,1\,1\,1\,1\,0\,0\,0\,0\,0\,0\,1\,0\,1 \\
1\,1\,1\,1\,1\,0\,1\,1\,1\,1\,0\,1\,0\,1\,1 \\
\vdots\,\vdots\,\vdots\,\vdots\,\vdots\,\vdots\,\vdots\,\vdots\,\vdots\,\vdots\,\vdots\,\vdots\,\vdots\,\vdots\,\vdots \\
1\,0\,0\,0\,0\,0\,1\,0\,0\,0\,0\,0\,0\,0\,0 \\
1\,1\,0\,0\,0\,0\,0\,0\,0\,0\,0\,0\,0\,0\,0 \\
1\,0\,0\,0\,0\,0\,0\,0\,0\,0\,0\,0\,0\,0\,0
\end{pmatrix}
$$

with

$$\mathcal{M}_2 = \text{PoissonRegressionModel}\,[X_2]$$

TABLE 7.1:   Lizard data: Estimates and
approximate 0.95-level confidence intervals of model
parameters for the model with two factor interactions.

Parameter	Estimate	Lower limit	Upper limit
β_s	0.734	0.0918	1.401
β_i	-1.581	-2.402	-0.812
β_d	0.481	-0.105	1.078
β_h	1.645	1.046	2.282
β_{tE}	0.0669	-0.644	0.786
β_{tM}	0.667	0.00718	1.347
β_{s*i}	0.835	0.232	1.493
β_{s*d}	0.749	0.337	1.165
β_{s*h}	-1.127	-1.650	-0.640
β_{s*tE}	0.752	0.165	1.342
β_{s*tM}	0.955	0.402	1.507
β_{i*d}	0.222	-0.260	0.718
β_{i*h}	0.0447	-0.425	0.523
β_{i*tE}	-0.167	-0.715	0.388
β_{i*tM}	-1.774	-2.411	-1.154
β_{d*h}	-0.607	-0.995	-0.226
β_{d*tE}	-0.106	-0.650	0.428
β_{d*tM}	-0.366	-0.883	0.138
β_{h*tE}	0.162	-0.370	0.692
β_{h*tM}	0.389	-0.127	0.902

defines the log-linear regression model with effects upto two factor interactions
without interaction of illumination and perch diameter, illumination and perch
height, perch diameter and time of day, and perch height and time of day, that
is, the model includes in addition to constant and main effects also interactions
of species with all other factors, illumination and time of day, and perch
diameter and perch height.

Table 7.2 gives MLEs and approximate 0.95-level profile likelihood-based
confidence intervals of model parameters.

The approximate 0.95-level profile likelihood-based confidence interval for
the difference of β_{i*tE} and β_{i*tM} is (1.086, 2.180). The estimate of $e^{\beta_{s*d}}$ and
its approximate 0.95-level profile likelihood-based confidence interval is 2.387
and (1.330,4.539), respectively. This can be interpreted also to mean that
odds for *Grahami* versus *Opalinus* are 2.387 higher in *Shade* than that in
Sun.

Figure 7.6 gives plots of residuals.

☐

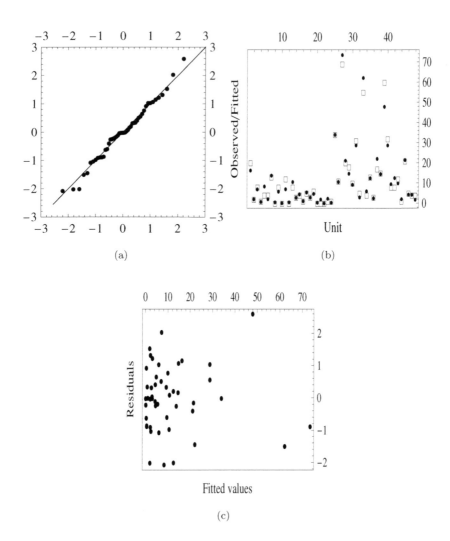

FIGURE 7.6: Lizard data: (a) QQ-plot of modified residuals, (b) scatter plot of observations (□) and their fitted values (•), and (c) scatter plot of modified residuals against fitted values.

TABLE 7.2: Lizard data: Estimates and approximate 0.95-level confidence intervals of model parameters for the model with selected two factor interactions.

Parameter	Estimate	Lower limit	Upper limit
β_{s}	0.812	0.181	1.473
β_{i}	-1.430	-2.095	-0.827
β_{d}	0.325	-0.141	0.795
β_{h}	1.882	1.402	2.405
β_{tE}	0.145	-0.334	0.630
β_{tM}	0.822	0.396	1.269
$\beta_{\mathrm{s*i}}$	0.870	0.285	1.513
$\beta_{\mathrm{s*d}}$	0.733	0.330	1.140
$\beta_{\mathrm{s*h}}$	-1.084	-1.600	-0.605
$\beta_{\mathrm{s*tE}}$	0.688	0.125	1.252
$\beta_{\mathrm{s*tM}}$	0.780	0.254	1.302
$\beta_{\mathrm{i*tE}}$	-0.172	-0.719	0.382
$\beta_{\mathrm{i*tM}}$	-1.790	-2.424	-1.174
$\beta_{\mathrm{d*h}}$	-0.631	-1.017	-0.252

7.6   Bibliographic notes

Lindsey (1995a) discusses Poisson regression and Haberman (1978) and Christensen (1990) are books on log-linear models.

Chapter 8

Multinomial Regression Model

8.1 Data

Suppose that a response z with k possible values denoted by integers $1, 2, \ldots,$ k has been observed from every statistical unit. The probability of outcome j of the statistical unit l is

$$\Pr(z_l = j) = \omega_{lj} \; \forall l = 1, \ldots, N, \text{ and } j = 1, \ldots, k,$$

such that

$$\sum_{j=1}^{k} \omega_{lj} = 1 \; \forall l = 1, \ldots, N.$$

Usually the probability of the jth outcome of the lth statistical unit is assumed to depend on the values $(x_{l1}, \ldots, x_{lq})$, of some factors associated with the unit, that is,

$$\Pr(z_l = j) = \omega_{lj} = \omega_j(x_{l1}, \ldots, x_{lq}) \; \forall l = 1, \ldots, N, \text{ and } j = 1, \ldots, k,$$

where the ω_js are given functions of factors $x_1, \ldots, x_q$.

Let there be n different combinations $(x_{i1}, \ldots, x_{iq}) \; \forall i = 1, \ldots, n$ of values of factors and let there be m_i statistical units with the combination $(x_{i1}, \ldots, x_{iq})$ of factors, that is, m_i statistical units in the data have the same combination $(x_{i1}, \ldots, x_{iq})$ of values of factors. These units are said to form a *covariate class*.

When in the data the number of covariate classes is considerably smaller than the number of statistical units, it is practical to generate new variables $y_{ij} \; \forall i = 1, \ldots, n$ and $\forall j = 1, \ldots, k$, where y_{ij} is the total number of outcomes j in the ith covariate class. Consider as new responses vectors

$$y_i = (y_{i1}, \ldots, y_{ik}) \; \forall i = 1, \ldots, n.$$

These new responses are statistically independent and

$$y_i \sim \text{MultinomialModel}[k, m_i] \; \forall i = 1, \ldots, n \tag{8.1}$$

with

$$\omega_i = (\omega_1(x_{i1}, \ldots, x_{iq}), \ldots, \omega_k(x_{i1}, \ldots, x_{iq})) \; \forall i = 1, \ldots, n$$

as the vector of probabilities of different outcomes of the ith covariate class. The original or ungrouped data can be considered to be a special case, where

$$m_1 = m_2 = \cdots = m_N = 1.$$

Assume now that the functions $w_j(x_1, \ldots, x_q)$ in addition to factors depends on d unknown common parameters $\beta_1, \ldots, \beta_d$, that is,

$$w_{ij}(\beta_1, \ldots, \beta_d) = w_j(x_{i1}, \ldots, x_{iq}; \beta_1, \ldots, \beta_d) \; \forall i = 1, \ldots, n \text{ and } j = 1, \ldots, k.$$
(8.2)

Then the model is a GRM. We call it *multinomial regression model* in general. This includes cases where success probabilities depend nonlinearly on regression parameters.

Example 8.1 Smoking data (cont'd)

Assuming that counts in each row of Table 2.5 in Chapter 2 can be modeled as observations from some multinomial distributions, which may have different probabilities of outcomes "*Nonsmoker*," "*Ex-smoker*," "*Occasional*," and "*Regular*," the appropriate model matrix is

$$X = \begin{pmatrix} 1 & 0 & 0 & 0 \\ 0 & 1 & 0 & 0 \\ 0 & 0 & 1 & 0 \\ 0 & 0 & 0 & 1 \end{pmatrix}.$$

The statistical model for the responses is defined by

$$\mathcal{M} = \text{MultinomialRegressionModel}[X, \{540, 212, 403, 529\}],$$

where the argument $\{540, 212, 403, 529\}$ consists of row totals, that is, numbers of trials at levels "*Neither*," "*Mother*," "*Father*," and "*Both*" of the parental behavior factor. The parameter of the model has the form

$$\theta = (\beta_{11}, \beta_{12}, \beta_{13}, \beta_{21}, \beta_{22}, \beta_{23}, \beta_{31}, \beta_{32}, \beta_{33}, \beta_{41}, \beta_{42}, \beta_{43}),$$

where the component β_{ij} denotes

$$\beta_{ij} = \ln\left(\frac{w_{ij}}{w_{i4}}\right) \; \forall i = 1, 2, 3, 4; j = 1, 2, 3,$$

and w_{ij} is the probability of outcome j in the row i, that is, log-odds of outcome j to outcome 4 for the row i. ⬜

8.2 Multinomial distribution

The moment generating function of MultinomialModel$[k, m]$ has the form

$$M_{(y_1, y_2, \ldots, y_k)}(\xi_1, \xi_2, \ldots, \xi_k) = \mathrm{E}(e^{\xi_1 y_1 + \xi_2 y_2 + \cdots + \xi_k y_k}) = \left(\sum_{j=1}^{k} w_j e^{\xi_j} \right)^m \quad (8.3)$$

and its cumulant generating function is

$$K_{(y_1, y_2, \ldots, y_k)}(\xi_1, \xi_2, \ldots, \xi_k) = m \ln \left(\sum_{j=1}^{k} w_j e^{\xi_j} \right). \quad (8.4)$$

Cumulants up to the order of four of the MultinomialModel$[k, m]$ are

$$\kappa_i = m w_i, \quad (8.5)$$
$$\kappa_{ii} = m w_i (1 - w_i), \quad (8.6)$$
$$\kappa_{ij} = -m w_i w_j \qquad \forall i \neq j, \quad (8.7)$$
$$\kappa_{iii} = m w_i (1 - w_i)(1 - 2w_i), \quad (8.8)$$
$$\kappa_{iij} = -m w_i (1 - 2w_i) w_j \qquad \forall i \neq j, \quad (8.9)$$
$$\kappa_{ijl} = 2 m w_j w_i w_l \qquad \forall i \neq j \neq l, \quad (8.10)$$
$$\kappa_{iiii} = m w_i (1 - w_i) \{1 - 6 w_i (1 - w_i)\}, \quad (8.11)$$
$$\kappa_{iiij} = -m w_i \{1 - 6 w_i (1 - w_i)\} w_j \qquad \forall i \neq j,, \quad (8.12)$$
$$\kappa_{iijj} = -m w_i w_j (1 - 2w_i - 2w_j + 6 w_i w_j), \qquad \forall i \neq j, \quad (8.13)$$
$$\kappa_{iijl} = 2 m w_i w_j w_l (1 - 3w_i) \qquad \forall i \neq j \neq l, \quad (8.14)$$
$$\kappa_{ijlr} = -6 m w_i w_j w_l w_r \qquad \forall i \neq j \neq l \neq r. \quad (8.15)$$

8.3 Likelihood function

When responses $y_1, \ldots, y_n$ are assumed to be statistically independent observations from multinomial distributions with numbers of trials m_i and success probabilities w_{ij}, the log-likelihood function has the form

$$\ell(w; y) = \sum_{i=1}^{n} \sum_{j=1}^{k} y_{ij} \ln(w_{ij}), \quad (8.16)$$

where

$$w = (w_1, \ldots, w_n) = ((w_{11}, w_{12}, \ldots, w_{1k}), \ldots, (w_{n1}, w_{n2}, \ldots, w_{nk}))$$

and

$$y = (y_1, y_2, \ldots, y_n) = ((y_{11}, y_{12}, \ldots, y_{1k}), \ldots, (y_{n1}, y_{n2}, \ldots, y_{nk})).$$

Relation between factors and vector ω

$$g(\omega_{ij}) = \eta_{ij} = \sum_{l=1}^{d} x_{il}\beta_{jl} \ \forall j = 1, \ldots, k-1, \forall i = 1, \ldots, n$$

results into a formula containing parameters $\beta_{11}, \beta_{12}, \ldots, \beta_{(k-1)d}$.

In the case of a logistic model

$$g(\omega_{ij}) = \eta_{ij} = \ln\left(\frac{\omega_{ij}}{\omega_{ik}}\right) = \sum_{l=1}^{d} x_{il}\beta_{jl} \ \forall j = 1, \ldots, k-1, \forall i = 1, \ldots, n.$$

The log-likelihood function now has the form

$$\ell(\beta; y) = \sum_{i=1}^{n} \left(\sum_{j=1}^{k-1} y_{ij} \sum_{l=1}^{d} x_{il}\beta_{jl} - m_i \ln\left(1 + \sum_{j=1}^{k-1} e^{\sum_{l=1}^{d} x_{il}\beta_{jl}} \right) \right)$$

$$= \sum_{i=1}^{n} \sum_{j=1}^{k-1} \sum_{l=1}^{d} y_{ij} x_{il}\beta_{jl} - \sum_{i=1}^{n} m_i \ln\left(1 + \sum_{j=1}^{k-1} e^{\sum_{l=1}^{d} x_{il}\beta_{jl}} \right),$$

which can be written in the form

$$\ell(\beta; y) = \sum_{j=1}^{k-1} \sum_{l=1}^{d} \left(\sum_{i=1}^{n} y_{ij} x_{il} \right) \beta_{jl} - \sum_{i=1}^{n} m_i \ln\left(1 + \sum_{j=1}^{k-1} e^{\sum_{l=1}^{d} x_{il}\beta_{jl}} \right).$$

Thus, the statistics z

$$\sum_{i=1}^{n} y_{ij} x_{il} \ \forall l = 1, \ldots, d; \ \forall j = 1, \ldots, k-1$$

together are minimal sufficient statistics for the parameters $\beta_{11}, \beta_{12}, \ldots, \beta_{(k-1)d}$.

Taking partial derivatives of the logarithmic likelihood function with respect to components of the parameter vector, one gets the maximum likelihood equations

$$\sum_{i=1}^{n} y_{ij} x_{il} - \sum_{i=1}^{n} m_i \omega_{ij} x_{il} = 0 \ \forall l = 1, \ldots, d; \ \forall j = 1, \ldots, k-1,$$

where the ω_{ij}s depend on parameters through

$$\omega_{ij} = \frac{e^{\sum_{l=1}^{d} x_{il}\beta_{jl}}}{1 + \sum_{r=1}^{k-1} e^{\sum_{l=1}^{d} x_{il}\beta_{rl}}} \ \forall j = 1, \ldots, k-1, \ \forall i = 1, \ldots, n.$$

In matrix form

$$X^T y = X^T \mu$$

with μ denoting the matrix of cell means expressed as functions of parameters, that is,

$$\mu_{ij} = m_i \frac{e^{\sum_{l=1}^{d} x_{il}\beta_{jl}}}{1 + \sum_{r=1}^{k-1} e^{\sum_{l=1}^{d} x_{il}\beta_{rl}}} \quad \forall j = 1, \ldots, k-1, \ \forall i = 1, \ldots, n$$

and

$$\mu_{ik} = m_i \frac{1}{1 + \sum_{r=1}^{k-1} e^{\sum_{l=1}^{d} x_{il}\beta_{rl}}} \quad \forall i = 1, \ldots, n.$$

8.4 Logistic multinomial regression model

Example 8.2 Smoking data (cont'd)

Now

$$X^T y = \begin{pmatrix} 1\,0\,0\,0 \\ 0\,1\,0\,0 \\ 0\,0\,1\,0 \\ 0\,0\,0\,1 \end{pmatrix}^T \begin{pmatrix} 279\ 79\ 72 \\ 109\ 29\ 10 \\ 192\ 48\ 32 \\ 252\ 65\ 32 \end{pmatrix} = \begin{pmatrix} 279\ 79\ 72 \\ 109\ 29\ 10 \\ 192\ 48\ 32 \\ 252\ 65\ 32 \end{pmatrix}$$

and the maximum likelihood equations in this case take the simple form

$$y = \mu,$$

where

$$\mu_{ij} = m_i \frac{e^{\beta_{ji}}}{1 + \sum_{r=1}^{3} e^{\beta_{ri}}} \quad \forall j = 1, \ldots, 3, \forall i = 1, \ldots, 4$$

and

$$\mu_{i4} = m_i \frac{1}{1 + \sum_{r=1}^{3} e^{\beta_{ri}}} \quad \forall i = 1, \ldots, 4.$$

So the maximum likelihood equations are

$$y_{ij} = m_i \frac{e^{\beta_{ji}}}{1 + \sum_{r=1}^{3} e^{\beta_{ri}}} \quad \forall j = 1, \ldots, 3, \forall i = 1, \ldots, 4.$$

It is easily found that their solution is simply

$$\beta_{ij} = \ln \left(\frac{y_{ij}}{y_{i4}} \right) \quad \forall i = 1, 2, 3, 4; j = 1, 2, 3.$$

TABLE 8.1:   Smoking data: Estimates and
approximate 0.95-level confidence intervals of model
parameters.

Parameter	Estimate	Lower limit	Upper limit
β_{11}	0.931	0.7133	1.155
β_{12}	-0.331	-0.623	-0.0438
β_{13}	-0.424	-0.725	-0.129
β_{21}	0.532	0.227	0.846
β_{22}	-0.792	-1.244	-0.363
β_{23}	-1.856	-2.585	-1.237
β_{31}	0.382	0.1615	0.606
β_{32}	-1.004	-1.344	-0.681
β_{33}	-1.409	-1.812	-1.037
β_{41}	0.336	0.1461	0.529
β_{42}	-1.019	-1.309	-0.741
β_{43}	-1.727	-2.121	-1.366

Table 8.1 gives estimates and approximate 0.95-level profile likelihood-based
confidence intervals of model parameters. Considered separately all the parameters are significant at 0.05-level, but that is of minor interest here. Of
greater interest is to compare corresponding parameters for different parental
behaviors. So consider first the statistical hypothesis

$$H : \beta_{1j} = \beta_{2j} = \beta_{3j} = \beta_{4j} \; \forall j = 1, 2, 3,$$

which means that smoking behavior would be the same for all groups. The
observed significance level is 6.715×10^{-7}, that is, there is strong statistical
evidence for the fact that smoking habits depend on those of the parents.
In fact looking more carefully at Figure 3.6 and Table 8.1, the difference
seems to follow from the first row, namely in the case of neither parents
smoking, the relative frequency of occasional smokers is high and regular
smokers low compared to other cases of parental smoking behavior. The
observed significance level of the hypothesis

$$H : \beta_{2j} = \beta_{3j} = \beta_{4j} \; \forall j = 1, 2, 3$$

is 0.667. Table 8.2 gives estimates and approximate 0.95-level profile likelihood-
based confidence intervals of model parameters in the submodel in which the
last three parental smoking behavior groups have identical distributions. Additionally, ratios of probabilities of occasional smoking and probabilities of
regular smoking are given. The interest functions have the following form:

$$\psi_1 = \frac{e^{\beta_3 - \gamma_3}(1 + e^{\gamma_1} + e^{\gamma_2} + e^{\gamma_3})}{1 + e^{\beta_1} + e^{\beta_2} + e^{\beta_3}}$$

and

$$\psi_2 = \frac{1 + e^{\gamma_1} + e^{\gamma_2} + e^{\gamma_3}}{1 + e^{\beta_1} + e^{\beta_2} + e^{\beta_3}},$$

TABLE 8.2:   Smoking data: Estimates and
approximate 0.95-level confidence intervals of interest
functions in the submodel.

Function	Estimate	Lower limit	Upper limit
β_1	0.931	0.713	1.155
β_2	-0.331	-0.623	-0.0438
β_3	-0.424	-0.725	-0.129
γ_1	0.388	0.258	0.520
γ_2	-0.971	-1.167	-0.781
γ_3	-1.623	-1.879	-1.380
ψ_1	2.061	1.513	2.806
ψ_2	0.621	0.513	0.745

where β_1, β_2, and β_3 denote parameters of the first group and γ_1, γ_2, and γ_3
denote common parameters of the last three groups. Probability to be an oc-
casional smoker when both parents are nonsmokers is twice the corresponding
probability in other groups and, similarly, the probability of being a regular
smoker is about 0.6 times that of other groups. Figure 8.1 gives graphs of
log-likelihood functions of parameters and interest functions.   ⬚

8.5   Proportional odds regression model

In the previous discussion even though possible outcomes were denoted by
integers 1, 2, ..., and k no ordering was supposed among the possible outcomes. Now the situation is assumed to be different so that the set of possible
outcomes is ordered such that the usual ordering of the integers reflects the
ordering of possible outcomes. Assume that data has been classified into
covariate classes and denote by

$$\gamma_{ij} = \omega_{i1} + \ldots + \omega_{ij} \ \forall i = 1, \ldots, n, \quad \text{and} \quad j = 1, \ldots, k \qquad (8.17)$$

the cumulative probabilities that satisfy the relations

$$0 \leq \gamma_{i1} \leq \gamma_{i2} \leq \ldots \leq \gamma_{i(k-1)} \leq \gamma_{ik} = 1 \ \forall i = 1, \ldots, n.$$

Cell probabilities can be expressed in terms of cumulative probabilities

$$\omega_{ij} = \gamma_{ij} - \gamma_{i(j-1)} \ \forall i = 1, \ldots, n, \quad \text{and} \quad j = 1, \ldots, k \qquad (8.18)$$

with the understanding that

$$\gamma_{i0} = 0 \quad \text{and} \quad \gamma_{ik} = 1 \ \forall i = 1, \ldots, n.$$

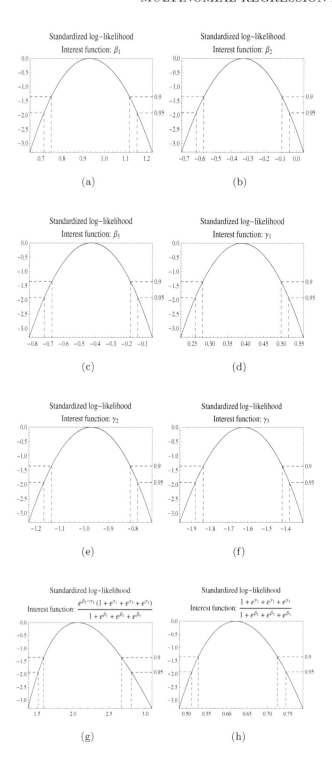

FIGURE 8.1: Smoking data: Plots of log-likelihoods of interest functions (a) β_1, (b) β_2, (c) β_3, (d) γ_1, (e) γ_2, (f) γ_3, (g) ψ_1, and (h) ψ_2.

Thus, use of cumulative probabilities is just a reparametrization of the multinomial model. This reparametrization is, however, better for taking account of the ordering of possible outcomes.

As above, suppose that for covariate class i the cumulative probabilities depend on the values $(x_{i1}, \ldots, x_{id})$ of factors associated with covariate class i, that is,

$$\gamma_{ij} = \gamma_j(x_{i1}, \ldots, x_{id}) \ \forall i = 1, \ldots, n, \text{ and } j = 1, \ldots, k. \tag{8.19}$$

Often the dependence of cumulative probabilities on factors occurs through the linear predictor $x_i^T \beta$ as follows:

$$\gamma_{ij} = \gamma_j(x_i^T \beta) \ \forall i = 1, \ldots, n, \text{ and } j = 1, \ldots, k. \tag{8.20}$$

For any continuous distribution function F and a set of intervals

$$(-\infty, \zeta_1], (\zeta_1, \zeta_2], \ldots, (\zeta_{k-2}, \zeta_{k-1}], (\zeta_{k-1}, \infty)$$

where

$$-\infty = \zeta_0 < \zeta_1 < \zeta_2 < \ldots < \zeta_{k-2} < \zeta_{k-1} < \zeta_k = \infty,$$

a model taking into account the ordering of all possible outcomes is obtained by defining

$$\gamma_{ij} = \gamma_j(x_i^T \beta) = F(\zeta_j - x_i^T \beta) \ \forall i = 1, \ldots, n, \text{ and } j = 1, \ldots, k. \tag{8.21}$$

Cell probabilities now have the form

$$\omega_{ij} = F(\zeta_j - x_i^T \beta) - F(\zeta_{j-1} - x_i^T \beta) \ \forall i = 1, \ldots, n, \text{ and } j = 1, \ldots, k. \tag{8.22}$$

Because probabilities of the first and last cells have the expressions

$$\omega_{i1} = F(\zeta_1 - x_i^T \beta) \text{ and } \omega_{ik} = 1 - F(\zeta_{k-1} - x_i^T \beta) \ \forall i = 1, \ldots, n$$

these probabilities decrease and increase, respectively, whenever $x_i^T \beta$ increases. Probabilities of cells with low index and high index tend to behave in a similar manner, but the situation with respect to these cell probabilities is more complicated because of relations between cell probabilities.

An important special case arises when F is selected to be the cumulative distribution of the standardized logistic distribution, that is,

$$F(z) = \frac{e^z}{1 + e^z} \ \forall -\infty < z < \infty.$$

For this case it is true that the odds ratios for two covariate classes x and $\tilde{x}$ has the form

$$\frac{\gamma_j(x^T \beta)/(1 - \gamma_j(x^T \beta))}{\gamma_j(\tilde{x}^T \beta)/(1 - \gamma_j(\tilde{x}^T \beta))} = \frac{e^{\zeta_j - x^T \beta}}{e^{\zeta_j - \tilde{x}^T \beta}} = e^{(\tilde{x} - x)^T \beta} \tag{8.23}$$

which is independent of j. Thus, all the odds ratios for covariate class x are proportional to the corresponding odds ratios for the covariate class $\tilde{x}$ with the same multiplier for all possible outcomes.

The logarithmic likelihood function now has the following form:

$$\ell(\omega; y) = \sum_{i=1}^{n} \sum_{j=1}^{k} y_{ij} \ln(F(\zeta_j - x_i^T \beta) - F(\zeta_{j-1} - x_i^T \beta)). \qquad (8.24)$$

Example 8.3 Coal miners data (cont'd)

The log-likelihood function calculated from statistical evidence is

$$\sum_{i=1}^{8} \left(y_{i1} \ln \left(\frac{1}{1 + e^{x_i \beta - \zeta_1}} \right) \right.$$

$$+ y_{i2} \ln \left(\frac{1}{1 + e^{x_i \beta - \zeta_2}} - \frac{1}{1 + e^{x_i \beta - \zeta_1}} \right)$$

$$\left. + y_{i3} \ln \left(1 - \frac{1}{1 + e^{x_i \beta - \zeta_2}} \right) \right).$$

The MLE is $(9.676, 10.582, 2.597)$ and estimated cell probabilities are

$$\begin{pmatrix} 0.994 & 0.00356 & 0.00243 \\ 0.934 & 0.0384 & 0.0279 \\ 0.847 & 0.0851 & 0.0682 \\ 0.745 & 0.134 & 0.122 \\ 0.636 & 0.176 & 0.188 \\ 0.532 & 0.206 & 0.262 \\ 0.434 & 0.221 & 0.345 \\ 0.364 & 0.222 & 0.414 \end{pmatrix}.$$

For comparison, observed cell probabilities are

$$\begin{pmatrix} 1. & 0. & 0. \\ 0.944 & 0.0370 & 0.0185 \\ 0.791 & 0.140 & 0.0698 \\ 0.729 & 0.104 & 0.167 \\ 0.627 & 0.196 & 0.176 \\ 0.605 & 0.184 & 0.211 \\ 0.429 & 0.214 & 0.357 \\ 0.364 & 0.182 & 0.455 \end{pmatrix}.$$

Table 8.3 gives estimates and approximate 0.95-level profile likelihood-based confidence intervals of model parameters and the following interest functions:

$$\psi_1 = 1 - \frac{1}{1 + e^{\beta \ln(25) - \zeta_2}}$$

TABLE 8.3: Coal miners data: Estimates and approximate 0.95-level confidence intervals of interest functions.

Function	Estimate	Lower limit	Upper limit
ζ_1	9.676	7.300	12.491
ζ_2	10.582	8.165	13.436
β	2.597	1.907	3.402
ψ_1	0.0977	0.0665	0.136
ψ_2	41.519	36.978	48.866

and

$$\psi_2 = e^{\frac{\zeta_1}{\beta}}.$$

The first interest function is the probability of severe pneumoconiosis at 25 years spent in a coal mine and the second is the time spent when the probability for being healthy is 0.5. Figure 8.2 gives graphs of log-likelihood functions of parameters and interest functions. ⬜

8.6   Bibliographic notes

Agresti (1990) discusses analysis of categorical data. The article by McCullagh (1980) introduced the proportional odds model and Agresti (1984) is a book on the subject.

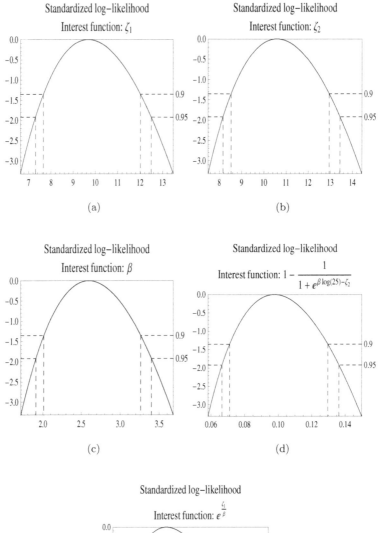

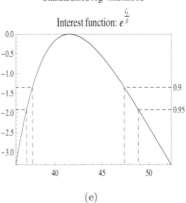

FIGURE 8.2: Coal miners data: Plots of log-likelihoods of interest functions (a) ζ_1, (b) ζ_2, (c) β, (d) ψ_1, and (e) ψ_2.

Chapter 9

Other Generalized Linear Regressions Models

9.1 Negative binomial regression model

9.1.1 Data

Suppose that a binary response z, the two possible values of which are denoted by 0 and 1, has been observed from every statistical unit. The probabilities of "failure" (0) and "success" (1) of the statistical unit j are

$$\Pr(z_j = 0) = 1 - \omega_j \text{ and } \Pr(z_j = 1) = \omega_j \ \forall j = 1, \ldots, N.$$

Usually the success probability of the jth statistical unit is assumed to depend on the values $(x_{j1}, \ldots, x_{jq})$ of the factors associated with the unit, that is,

$$\Pr(z_j = 1) = \omega_j = \omega(x_{j1}, \ldots, x_{jq}) \ \forall j = 1, \ldots, N,$$

where ω is a given function of factors $x_1, \ldots, x_q$. Let there be n different combinations $(x_{i1}, \ldots, x_{iq}) \ \forall i = 1, \ldots, n$ of values of factors and let there be m_i and y_i statistical units with the combination $(x_{i1}, \ldots, x_{iq})$ of factors such that their observation was a success and failure, respectively. These units are said to form a covariate class.

Now suppose that numbers of successes in covariance classes are fixed and numbers of failures in them are random. Then assuming independence among observations in a covariate class the distribution of number of failures is

$$y_i \sim \text{NegativeBinomialModel}[m_i, \omega_i] \ \forall i = 1, \ldots, n, \tag{9.1}$$

where

$$\omega_i = \omega(x_{i1}, \ldots, x_{iq}) \ \forall i = 1, \ldots, n \tag{9.2}$$

is the success probability of the ith covariate class. In the process of aggregation of the data into the counts of failures in covariate classes information may be lost.

Assume now that the function $\omega(x_1, \ldots, x_q)$ in addition to factors depends on d unknown common parameters $\beta_1, \ldots, \beta_d$, that is,

$$\omega_i(\beta_1, \ldots, \beta_d) = \omega(x_{i1}, \ldots, x_{iq}; \beta_1, \ldots, \beta_d) \ \forall i = 1, \ldots, n. \tag{9.3}$$

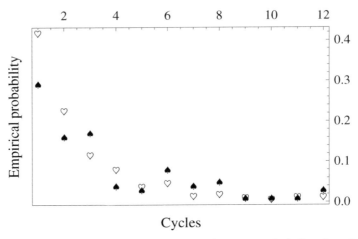

FIGURE 9.1: Pregnancy data: Observed point probability functions for smokers (♠) and nonsmokers (♡).

Then the model is a GRM. We call it a *negative binomial regression model* in general. This includes cases where success probabilities depend nonlinearly on regression parameters. If the function ω has the form

$$\omega(x_{i1}, \ldots, x_{id}; \beta_1, \ldots, \beta_d) = h(x_{i1}\beta_1 + x_{i2}\beta_2 + \cdots + x_{id}\beta_d) \; \forall i = 1, \ldots, n \quad (9.4)$$

for some known function h, the model is a GLIM. The function h is the inverse of the link function, that is, $g = h^{-1}$. Finally, when $h(z) = \frac{e^z}{1+e^z}$ the model is called a *negative binomial logistic regression model*.

Link functions for binomial regression models presented in Section 6.3 also apply to negative binomial regression models.

Example 9.1 Pregnancy data

A total of 678 women, who became pregnant under planned pregnancies, were asked how many cycles it took them to get pregnant. The women were classified as smokers and nonsmokers; it is of interest to compare the association between smoking and probability of pregnancy. Columns in Table A.16 in Appendix A on page 263 contain values of cycles, smoking habit, and count of women. Figure 9.1 shows a plot of empirical point probability functions for smokers and nonsmokers. This plot indicates that success probability is higher for nonsmokers than for smokers.

Assume that the data can be modeled as two samples from geometric distributions with unknown success probabilities. Then the response vector consists of 678 values so that for every woman the observation is value of cycle minus one except for 7 smokers and 12 nonsmokers for which it is only known that the value is at least 12, that is,

$$y_{\text{obs}} = (0, 0, 0, \ldots, [12, \infty), [12, \infty), [12, \infty))^{\text{T}}.$$

The model matrix generated by the smoking factor is

$$X = \begin{pmatrix} 1 & 1 \\ 1 & 1 \\ 1 & 1 \\ \vdots & \vdots \\ 1 & 0 \\ 1 & 0 \\ 1 & 0 \end{pmatrix}.$$

The statistical model with logit-link is defined by the following expression:

$$\mathcal{M} = \text{RegressionModel}\Big[X, \text{Distribution} \rightarrow \text{GeometricModel}[\omega],$$

$$\text{InverseLink} \rightarrow \frac{e^z}{1 + e^z}\Big].$$

▯

9.1.2   Negative binomial distribution

The moment generating function of NegativeBinomialModel$[m, \omega]$ is

$$M(\xi) = \left(\frac{\omega}{1 - (1 - \omega)e^\xi} \right)^m \tag{9.5}$$

and thus the cumulant generating function is

$$K(\xi) = m \ln \left(\frac{\omega}{1 - (1 - \omega)e^\xi} \right). \tag{9.6}$$

The first four cumulants of NegativeBinomialModel$[m, \omega]$ are

$$\kappa_1 = \frac{m(1 - \omega)}{\omega}, \tag{9.7}$$

$$\kappa_2 = \frac{m(1 - \omega)}{\omega^2}, \tag{9.8}$$

$$\kappa_3 = \frac{m(1 - \omega)(2 - \omega)}{\omega^3}, \tag{9.9}$$

$$\kappa_4 = \frac{m(1 - \omega)(6 - 6\omega + \omega^2)}{\omega^4}. \tag{9.10}$$

The cumulants of the standardized variable

$$z = \frac{y - \frac{m(1-\omega)}{\omega}}{\sqrt{\frac{m(1-\omega)}{\omega^2}}}$$

are seen from the following Taylor series expansion of the cumulant generating function

$$\frac{1}{2}\xi^2 + \frac{(2-\omega)}{6m^{1/2}(1-\omega)^{1/2}}\xi^3 + \frac{(\omega^2 - 6\omega + 6)}{24m(1-\omega)}\xi^4$$
$$- \frac{(\omega^3 - 14\omega^2 + 36\omega - 24)}{120m^{3/2}(1-\omega)^{3/2}}\xi^5$$
$$+ \frac{(\omega^4 - 30\omega^3 + 150\omega^2 - 240\omega + 120)}{720m^2(1-\omega)^2}\xi^6 + O(\xi^7)$$

to be of order 0, 1, $O(m^{-1/2})$, $O(m^{-1})$, etc. When $r \geq 2$, the rth cumulant of z is of order $O(m^{1-r/2})$. When ω is fixed, the cumulants of z approach the cumulants 0, 1, 0, 0, ... of standardized normal distribution as $m \to \infty$. Because convergence of cumulants implies convergence in distribution, the approximate probabilities will be obtained from the formulae

$$\Pr(y \geq \tilde{y}) \approx 1 - \Phi(z^-)$$

and

$$\Pr(y \leq \tilde{y}) \approx \Phi(z^+),$$

where Φ is the cumulative distribution function of the standardized normal distribution, $\tilde{y}$ is an integer, and

$$z^- = \frac{\tilde{y} - \frac{m(1-\omega)}{\omega} - \frac{1}{2}}{\sqrt{\frac{m(1-\omega)}{\omega^2}}} \quad \text{and} \quad z^+ = \frac{\tilde{y} - \frac{m(1-\omega)}{\omega} + \frac{1}{2}}{\sqrt{\frac{m(1-\omega)}{\omega^2}}}.$$

Thus, the affect of the continuity correction $\mp\frac{1}{2}$ on the probabilities is of order $O(m^{-1/2})$ and so asymptotically negligible. In medium size samples the effect of the continuity correction is remarkable and almost always improves the approximation.

Figure 9.2 contains graphs of point probability functions of negative binomial distributions with success probability 0.2 and numbers of successes 1, 2, 4, and 8.

9.1.3 Likelihood inference

Likelihood function of negative binomial data

When responses $y_1, \ldots, y_n$ are assumed to be statistically independent observations from negative binomial distributions with numbers of successes m_i

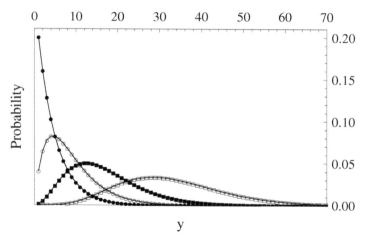

FIGURE 9.2: Point probability functions of negative binomial distributions with success probability 0.2 and number of successes 1 ($\bullet$), 2 ($\circ$), 4 ($\blacksquare$), and 8 ($\square$).

and success probabilities ω_i, the log-likelihood function has the form

$$\ell(\omega; y) = \sum_{i=1}^{n} \left[y_i \ln(1 - \omega_i) + m_i \ln(\omega_i) + \ln\binom{m_i + y_i - 1}{y_i} \right], \qquad (9.11)$$

where $\omega = (\omega_1, \omega_2, \ldots, \omega_n)$ and $y = (y_1, y_2, \ldots, y_n)$. Relation between factors and vector ω

$$g(\omega_i) = \eta_i = \sum_j x_{ij}\beta_j \ \forall i = 1, \ldots, n$$

results in a formula containing parameters $\beta_1, \beta_2, \ldots, \beta_d$.

In the case of link function

$$g(\omega_i) = \eta_i = \ln(1 - \omega_i) = \sum_j x_{ij}\beta_j \ \forall i = 1, \ldots, n$$

the log-likelihood function has the form

$$\ell(\beta; y) = \sum_{i=1}^{n} \left[y_i \sum_{j=1}^{d} x_{ij}\beta_j - m_i \ln\left(1 - e^{\sum_{j=1}^{d} x_{ij}\beta_j}\right) + \ln\binom{m_i + y_i - 1}{y_i} \right]$$

$$= \sum_{i=1}^{n} \sum_{j=1}^{d} y_i x_{ij}\beta_j - \sum_{i=1}^{n} m_i \ln\left(1 - e^{\sum_{j=1}^{d} x_{ij}\beta_j}\right) + \sum_{i=1}^{n} \ln\binom{m_i + y_i - 1}{y_i},$$

which can be written in the form

$$\ell(\beta; y) = \sum_{j=1}^{d} \beta_j \left(\sum_{i=1}^{n} y_i x_{ij} \right) - \sum_{i=1}^{n} m_i \ln \left(1 - e^{\sum_j x_{ij} \beta_j} \right) +$$

$$\sum_{i=1}^{n} \ln \left(\frac{m_i + y_i - 1}{y_i} \right).$$

Thus, the statistics

$$\sum_{i=1}^{n} y_i x_{ij} \ \forall j = 1, \ldots, d$$

together are a minimal sufficient statistic for the parameters $\beta_1, \beta_2, \ldots, \beta_d$.

The corresponding inverse link function has the form

$$h(\eta_i) = \omega_i = 1 - e^{\eta_i} = 1 - e^{-(-\eta_i)} \ \forall i = 1, \ldots, n.$$

This means that the inverse link function is the cumulative distribution function of a random variable, such that the negative of that random variable has an exponential distribution with mean 1. Use of this link function is less attractive because the linear predictor is constrained to take only negative values.

Example 9.2 Pregnancy data (cont'd)

The observed log-likelihood function has the following form:

$$566\beta_1 + 92\beta_2 - 1429 \ln(1 + e^{\beta_1}) - 409 \ln(1 + e^{\beta_1 + \beta_2}).$$

Figure 9.3 shows the contour plot of the log-likelihood function.   ⬚

Estimates of parameters

Because

$$\frac{\partial \ell}{\partial \omega_i} = \frac{m_i}{\omega_i} - \frac{y_i}{1 - \omega_i},$$

using the chain rule gives

$$\frac{\partial \ell}{\partial \beta_r} = \sum_{i=1}^{n} \left(\frac{m_i}{\omega_i} - \frac{y_i}{1 - \omega_i} \right) \frac{\partial \omega_i}{\partial \beta_r}.$$

Using the chain rule again we have

$$\frac{\partial \omega_i}{\partial \beta_r} = \frac{\partial \omega_i}{\partial \eta_i} \frac{\partial \eta_i}{\partial \beta_r} = \frac{\partial \omega_i}{\partial \eta_i} x_{ir},$$

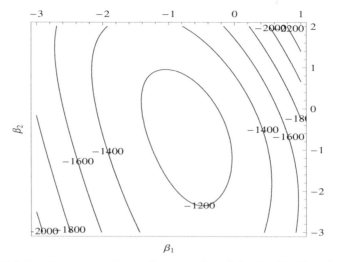

FIGURE 9.3:   Pregnancy data: Contour plot of the log-likelihood function.

and so

$$\frac{\partial \ell}{\partial \beta_r} = \sum_{i=1}^{n} \left(\frac{m_i}{\omega_i} - \frac{y_i}{1 - \omega_i} \right) \frac{\partial \omega_i}{\partial \eta_i} x_{ir}.$$

Fisher's expected information is

$$-\mathrm{E} \left(\frac{\partial^2 \ell}{\partial \beta_r \partial \beta_s} \right) = \sum_{i=1}^{n} \frac{m_i}{\omega_i^2 (1 - \omega_i)} \frac{\partial \omega_i}{\partial \beta_r} \frac{\partial \omega_i}{\partial \beta_s}$$

$$= \sum_{i=1}^{n} m_i \frac{\left(\frac{\partial \omega_i}{\partial \eta_i} \right)^2}{\omega_i^2 (1 - \omega_i)} x_{ir} x_{is} = \{X^{\mathrm{T}} W X\}_{rs},$$

where W is the diagonal matrix of weights

$$W = \mathrm{diag} \left\{ \frac{m_i \left(\frac{\partial \omega_i}{\partial \eta_i} \right)^2}{\omega_i^2 (1 - \omega_i)} \right\}.$$

Using this weight matrix in Equation (2.16) produces MLE.

Example 9.3 Pregnancy data (cont'd)

Maximum likelihood equations are

$$\begin{pmatrix} -1272 + \frac{1429}{1 + e^{\beta_1}} + \frac{409}{1 + e^{\beta_1 + \beta_2}} \\ -317 + \frac{409}{1 + e^{\beta_1 + \beta_2}} \end{pmatrix} = \begin{pmatrix} 0 \\ 0 \end{pmatrix}$$

and their solution is $(-0.701, -0.537)$. □

Likelihood ratio statistic or deviance function

In the fitted model the log-likelihood is

$$\ell(\hat{\omega}; y) = \sum_{i=1}^{n} \left[y_i \ln(1 - \hat{\omega}_i) + m_i \ln(\hat{\omega}_i) + \ln\binom{m_i + y_i - 1}{y_i} \right].$$

The values $\tilde{\omega}_i = m_i/(m_i + y_i)$ produce the highest likelihood. Thus, the deviance function is

$$D(y; \hat{\omega}) = 2\ell(\tilde{\omega}; y) - 2\ell(\hat{\omega}; y) = 2 \sum_{i=1}^{n} \left[y_i \ln\left(\frac{1 - \tilde{\omega}_i}{1 - \hat{\omega}_i}\right) + m_i \ln\left(\frac{\tilde{\omega}_i}{\hat{\omega}_i}\right) \right]. \quad (9.12)$$

The asymptotic distribution of the deviance $D(y; \hat{\omega})$ is χ^2-distribution with $n - d$ degrees of freedom under the following assumptions:

In the limit $D(y; \hat{\omega})$ is approximately independent of the estimated parameters $\hat{\beta}$ and thus approximately independent of the fitted probabilities $\hat{\omega}$. This independence is essential if D is used as a goodness of fit test statistic.

The deviance function is more useful for comparing nested models than as a measure of goodness of fit.

9.1.4   Negative binomial logistic regression model

The negative binomial logistic regression model is a negative binomial regression model such that the mean of the observation y_i can be written in the form

$$\mu_i = m_i e^{-(x_{i1}\beta_1 + x_{i2}\beta_2 + \cdots + x_{id}\beta_d)} \ \forall i = 1, \ldots, n$$

where m_i is the number of successes in the negative binomial distribution of y_i. Because the mean of the negative binomial distribution is equal to

$$\mu_i = m_i \frac{1 - \omega_i}{\omega_i}$$

this means that logit of the success probability of the negative binomial distribution is

$$\ln\left(\frac{\omega_i}{1 - \omega_i}\right) = \ln\left(\frac{m_i}{\mu_i}\right) = \eta_i = x_{i1}\beta_1 + x_{i2}\beta_2 + \cdots + x_{id}\beta_d \ \forall i = 1, \ldots, n.$$

Thus, the negative binomial logistic regression model is a generalized linear model with logit function as link function.

Example 9.4 Pregnancy data (cont'd)

In addition to model parameters assume that success probabilities for smokers and nonsmokers together with their differences and probabilities of no

TABLE 9.1: Pregnancy data: Estimates and approximate 0.95-level confidence intervals of interest functions.

Parameter function	Estimate	Lower limit	Upper limit
β_1	-0.701	-0.811	-0.591
β_2	-0.537	-0.798	-0.283
ψ_1	0.332	0.308	0.356
ψ_2	0.225	0.186	0.267
ψ_3	-0.107	-0.153	-0.0583
ψ_4	0.00794	0.00505	0.0121
ψ_5	0.0470	0.0240	0.0843
ψ_6	0.0391	0.0157	0.0765

success during 12 cycles for both groups with their differences are of interest. The interest functions have the following forms:

$$\psi_1 = \frac{e^{\beta_1}}{1 + e^{\beta_1}}$$

$$\psi_2 = \frac{e^{\beta_1+\beta_2}}{1 + e^{\beta_1+\beta_2}}$$

$$\psi_3 = \frac{e^{\beta_1+\beta_2}}{1 + e^{\beta_1+\beta_2}} - \frac{e^{\beta_1}}{1 + e^{\beta_1}}$$

$$\psi_4 = \left(1 - \frac{e^{\beta_1}}{1 + e^{\beta_1}}\right)^{12}$$

$$\psi_5 = \left(1 - \frac{e^{\beta_1+\beta_2}}{1 + e^{\beta_1+\beta_2}}\right)^{12}$$

$$\psi_6 = \left(1 - \frac{e^{\beta_1+\beta_2}}{1 + e^{\beta_1+\beta_2}}\right)^{12} - \left(1 - \frac{e^{\beta_1}}{1 + e^{\beta_1}}\right)^{12}$$

Table 9.1 gives MLEs and approximate 0.95-level profile likelihood-based confidence intervals of model parameters and interest functions and Figure 9.4 shows log-likelihood functions of parameters and interest functions. From Table 9.1 it is seen that the success probablity of nonsmokers is significantly greater than that of smokers. Consequently, the probability of no success during 12 cycles is considerably larger for smokers. According to the confidence interval, the difference of probabilities of no success might be as large as 0.0765. □

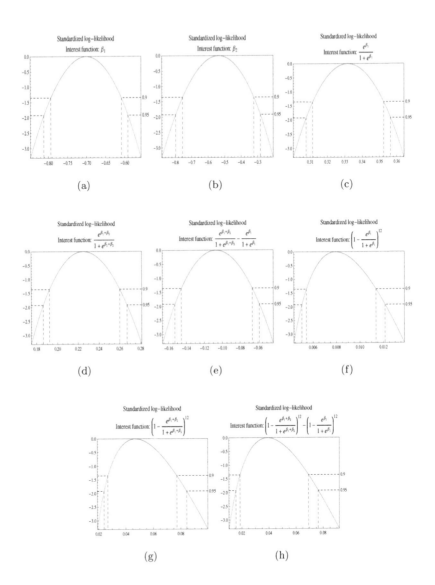

FIGURE 9.4:   Pregnancy data: Plots of log-likelihoods of interest functions
(a) β_1, (b) β_2, (c) ψ_1, (d) ψ_2, (e) ψ_3, (f) ψ_4, (g) ψ_5, and (h) ψ_6.

9.2   Gamma regression model

9.2.1   Data

Suppose that a response y_i has been observed for N statistical units. A basic assumption in this chapter is the following:

$$\text{var}(y_i) = \sigma^2 (\text{E}(y_i))^2 \ \forall i = 1, \ldots, N, \tag{9.13}$$

that is, the coefficient of variation of observations is constant and the common coefficient of variation is denoted by σ^2. If the possible values of responses consist of positive real numbers and if the responses are observations from gamma distributions, a special case, where $\sigma^2 = \frac{1}{\nu}$ and ν is the shape parameter, arises.

For the statistical unit i let

$$\text{E}(y_i) = \mu_i \ \forall i = 1, \ldots, N.$$

Usually the mean of a unit i is supposed to depend on the values $(x_{i1}, \ldots, x_{id})$ of factors connected with the unit, that is,

$$\text{E}(y_i) = \mu_i = \mu(x_{i1}, \ldots, x_{id}) \ \forall i = 1, \ldots, N, \tag{9.14}$$

where $\mu(x_1, \ldots, x_d)$ is a given function of the factors.

Example 9.5 Brain weights data

Table A.3 on page 254 in Appendix A shows mean weights of the brains and bodies of 62 species of mammals. The columns contain values of species, brain weight in grams, and body weight in kilograms. Figure 9.5 gives a graph of the natural logarithm of brain weight versus that of body weight showing clear linear relation between these quantities. Assume therefore that brain weights are observations from gamma distributions such that their means depend on logarithms of body weights.   ▯

9.2.2   Gamma distribution

The moment generating function of GammaModel$[\nu, \mu]$ has the form

$$M(\xi; \nu, \mu) = \left(1 - \frac{\xi\mu}{\nu}\right)^{-\nu} \tag{9.15}$$

and its cumulant generating function is

$$K(\xi) = \ln(M(\xi)) = -\nu \ln\left(1 - \frac{\xi\mu}{\nu}\right). \tag{9.16}$$

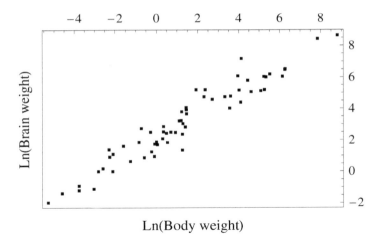

Ln(Body weight)

FIGURE 9.5: Brain weights data: Plot of logarithm of brain weight against logarithm of body weight.

The kth moment is

$$m_k = \frac{\mu^k(1+\nu)(2+\nu)\cdots(k-1+\nu)}{\nu^{k-1}} \quad \forall k = 1, 2, \ldots. \tag{9.17}$$

and the kth cumulant is

$$m_k = \frac{(k-1)!\mu^k}{\nu^{k-1}} \quad \forall k = 1, 2, \ldots. \tag{9.18}$$

Thus, the first four cumulants of the GammaModel$[\nu, \mu]$ are

$$\kappa_1 = \mu, \tag{9.19}$$

$$\kappa_2 = \frac{\mu^2}{\nu}, \tag{9.20}$$

$$\kappa_3 = \frac{2\mu^3}{\nu^2}, \tag{9.21}$$

$$\kappa_4 = \frac{6\mu^4}{\nu^3}. \tag{9.22}$$

The cumulants of the standardized variable

$$z = \frac{\sqrt{\nu}(y - \mu)}{\mu}$$

are seen from the Taylor series expansion

$$\frac{\xi^2}{2} + \frac{\xi^3}{3\sqrt{\nu}} + \frac{\xi^4}{4\nu} + \frac{\xi^5}{5\nu^{3/2}} + \frac{\xi^6}{6\nu^2} + O[\xi]^7$$

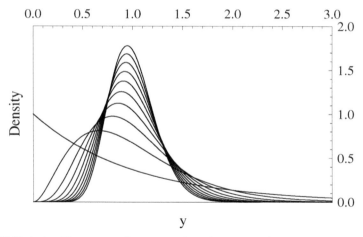

FIGURE 9.6: Densities of gamma distribution with mean 1 and shape 1, 3, ..., 19.

to be of order 0, 1, $O(\nu^{-1/2})$, $O(\nu^{-1})$, etc. When $r \geq 2$, the r^{th}s cumulant of z is of order $O(\nu^{1-r/2})$. The cumulants of Z approach the cumulants 0, 1, 0, 0, ... of the standardized normal distribution as $\nu \to \infty$. Because convergence of cumulants implies convergence in distribution, the approximate probabilities will be obtained from the formula

$$\Pr(Y \leq y) \approx \Phi\left(\frac{y - \mu}{\mu/\sqrt{\nu}}\right),$$

where Φ is the cumulative distribution function of the standardized normal distribution.

Figure 9.6 shows graphs of gamma densities with mean one and selected values of shape parameter ν.

9.2.3   Link function

The commonly used link function is the reciprocal link, that is,

$$g(\mu) = -\frac{1}{\mu}. \tag{9.23}$$

This link function is the canonical link. The reciprocal link function has the drawback that the linear predictor is restricted to take only negative values.

Given two distributions with cumulative distribution functions F_1 and F_2 such that the first distribution can have only positive values and the second can have as value any real number, then the function

$$g(\mu) = -F_2^{-1}(F_1(\mu)) \tag{9.24}$$

is a possible link function.

In the case of log-link the distributions are the standardized log-normal and normal distributions because

$$F_2^{-1}(F_1(\mu)) = \Phi^{-1}(\Phi(\ln(\mu))) = \ln(\mu).$$

For example, taking F_1 to be the cumulative distribution function of the ExpontialModel$[\mu]$ and F_2 the standardized normal distribution, we get

$$F_2^{-1}(F_1(\mu)) = \Phi^{-1}(1 - e^{-\mu}).$$

Figure 9.7 gives graphs of the previous three link functions separately and together.

Example 9.6 Brain weights data (cont'd)

Consider the gamma regression model with a log-link function with the model matrix

$$X = \begin{pmatrix} 1 & 1.219 \\ 1 & -0.734 \\ 1 & 0.300 \\ \vdots & \vdots \\ 1 & 0.482 \\ 1 & -2.263 \\ 1 & 1.443 \end{pmatrix}.$$

The gamma regression model is denoted by

$$\begin{aligned} \mathcal{M} = \ &\text{GeneralizedLinearRegressionModel}[X, \\ &\text{Distribution} \to \text{GammaModel}[\nu, \mu], \\ &\text{LinkFunction} \to \text{Log}]. \end{aligned}$$

□

9.2.4   Likelihood inference

Likelihood function of gamma data

When responses $y_1, \ldots, y_n$ are assumed to be statistically independent observations from gamma distributions with means μ_i and common shape ν, the log-likelihood function has the form

$$\ell(\mu, \nu; y) = \sum_{i=1}^{n} \left(-\frac{y_i \nu}{\mu_i} + (\nu - 1)\ln(y_i) + \nu \ln\left(\frac{\nu}{\mu_i}\right) - \ln(\Gamma(\nu)) \right), \quad (9.25)$$

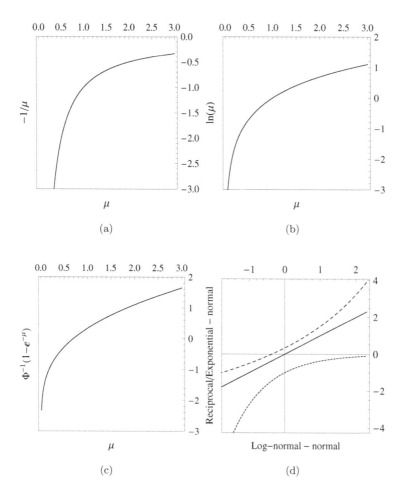

FIGURE 9.7:   Plots of link functions for gamma distribution: (a) reciprocal link, (b) log-link, that is, log-normal – normal link, (c) exponential – normal link, and (d) parametric plot of link functions.

where $\mu = (\mu_1, \mu_2, \ldots, \mu_n)$ and $y = (y_1, y_2, \ldots, y_n)$. Relation between factors and vector μ

$$g(\mu_i) = \eta_i = \sum_{j=1}^{d} x_{ij}\beta_j \ \forall i = 1, \ldots, n$$

results in a formula containing parameters $\beta_1, \beta_2, \ldots, \beta_d$.

In the case of the canonical link-function

$$g(\mu_i) = -\frac{1}{\mu_i} = \eta_i = \sum_{j=1}^{d} x_{ij}\beta_j \ \forall i = 1, \ldots, n.$$

The log-likelihood function now has the form

$$\ell(\beta, \nu; y) = \sum_{i=1}^{n}\left[\frac{y_i \sum_{j=1}^{d} x_{ij}\beta_j + \ln\left(-\sum_{j=1}^{d} x_{ij}\beta_j\right)}{1/\nu} + \right.$$

$$\left. (\nu - 1)\ln(y_i) + \nu\ln(\nu) - \ln(\Gamma(\nu))\right]$$

$$= \frac{\sum_{i=1}^{n}\sum_{j=1}^{d} y_i x_{ij}\beta_j + \sum_{i=1}^{n}\ln\left(-\sum_{j=1}^{d} x_{ij}\beta_j\right)}{1/\nu} +$$

$$\sum_{i=1}^{n}\left[(\nu - 1)\ln(y_i) + \nu\ln(\nu) - \ln(\Gamma(\nu))\right],$$

which can be written in the form

$$\ell(\beta, \nu; y) = \frac{\sum_{j=1}^{d}\left(\sum_{i=1}^{n} y_i x_{ij}\right)\beta_j + \sum_{i=1}^{n}\ln\left(-\sum_{j=1}^{d} x_{ij}\beta_j\right)}{1/\nu} +$$

$$\sum_{i=1}^{n}\left[(\nu - 1)\ln(y_i) + \nu\ln(\nu) - \ln(\Gamma(\nu))\right].$$

Thus, the statistics

$$\sum_{i=1}^{n} y_i x_{ij} \ \forall j = 1, \ldots, d$$

together are a minimal sufficient statistic for the parameters $\beta_1, \beta_2, \ldots, \beta_d$ for fixed ν.

Estimates of parameters

Because

$$\frac{\partial \ell}{\partial \mu_i} = \frac{\nu(y_i - \mu_i)}{\mu_i^2},$$

using the chain rule gives

$$\frac{\partial \ell}{\partial \beta_r} = \nu \sum_{i=1}^{n} \frac{(y_i - \mu_i)}{\mu_i^2} \frac{\partial \mu_i}{\partial \beta_r}.$$

Using the chain rule again we have

$$\frac{\partial \mu_i}{\partial \beta_r} = \frac{\partial \mu_i}{\partial \eta_i} \frac{\partial \eta_i}{\partial \beta_r} = \frac{\partial \mu_i}{\partial \eta_i} x_{ir},$$

and so

$$\frac{\partial \ell}{\partial \beta_r} = \nu \sum_{i=1}^{n} \frac{(y_i - \mu_i)}{\mu_i^2} \frac{\partial \mu_i}{\partial \eta_i} x_{ir}.$$

Fisher's expected information is

$$-\mathrm{E}\left(\frac{\partial^2 \ell}{\partial \beta_r \partial \beta_s}\right) = \sum_{i=1}^{n} \frac{1}{\mu_i^2} \frac{\partial \mu_i}{\partial \beta_r} \frac{\partial \mu_i}{\partial \beta_s} = \sum_{i=1}^{n} \frac{\left(\frac{\partial \mu_i}{\partial \eta_i}\right)^2}{\mu_i^2} x_{ir} x_{is} = \{X^{\mathrm{T}} W X\}_{rs},$$

where W is the diagonal matrix of weights

$$W = \mathrm{diag}\left\{\frac{\left(\frac{\partial \mu_i}{\partial \eta_i}\right)^2}{\mu_i^2}\right\}.$$

Using this weight matrix in Equation (2.16) produces MLE.

In the case of the canonical link-function we get

$$\frac{\partial \ell}{\partial \beta} = X^{\mathrm{T}}(y - \mu).$$

In addition, the diagonal matrix of weights has the form

$$W = \mathrm{diag}\{\mu_1^2, \ldots, \mu_n^2\}.$$

Example 9.7 Brain weights data (cont'd)

In addition to MLEs, Table 9.2 also contains approximate 0.95-level profile likelihood-based confidence intervals of parameters. Figure 9.8 shows graphs of the log-likelihood functions of parameters.

There exist explanations for the fact that the slope of the linear predictor is less than one. Those explanations predict that the slope should be three over four and the profile likelihood-based confidence interval for the slope is in agreement with this. Additionally, it is of interest to study the place of humans among mammals. For this purpose consider the interest functions consisting of mean brain weight at mean human body weight, the probability that brain

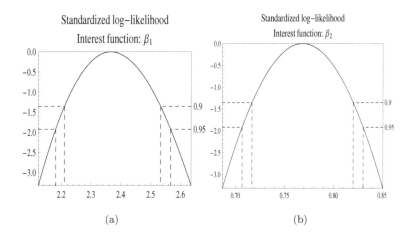

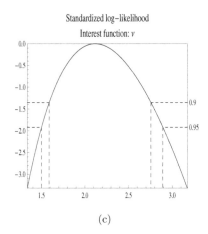

FIGURE 9.8:   Brain weights data: Plots of log-likelihoods of model parameters (a) β_1, (b) β_2, and (c) ν.

TABLE 9.2: Brain weights data: Estimates and approximate 0.95-level confidence intervals of model parameters.

Parameter	Estimate	Lower limit	Upper limit
β_1	2.367	2.181	2.566
β_2	0.768	0.707	0.830
ν	2.115	1.499	2.892

TABLE 9.3: Brain weights data: Estimates and approximate 0.95-level confidence intervals of interest functions.

Parameter function	Estimate	Lower limit	Upper limit
ψ_1	254.30	201.08	328.19
ψ_2	0.000258	0.00000837	0.00383
ψ_3	528.58	339.53	848.23

weight is at least that of observed human brain weight at mean human body weight, and predicted body weight for mammals with brain weight equal to the observed mean human brain weight. These interest functions are

$$\psi_1 = e^{\beta_1 + 4.1271\beta_2},$$

$$\psi_2 = Q\left(\nu, \frac{1320.02\nu}{\beta_1 + 4.1271\beta_2}\right),$$

and

$$\psi_3 = e^{\frac{7.1854 - \beta_1}{\beta_2}},$$

where Q is the so-called regularized gamma function. Table 9.3 gives estimates and approximate 0.95-level confidence intervals for interest functions. Figure 9.9 gives plots of log-likelihoods of interest functions. The log-likelihood of ψ_2 deviates so much from the quadratic form that the nominal confidence level in Table 9.3 is not reliable. □

Likelihood ratio statistic or deviance function

In a fitted model with fixed ν, the log-likelihood is

$$\ell(\hat{\mu}, \nu; y) = \sum_{i=1}^{n}\left(-\frac{y_i\nu}{\hat{\mu}_i} + (\nu - 1)\ln(y_i) + \nu\ln\left(\frac{\nu}{\hat{\mu}_i}\right) - \ln(\Gamma(\nu))\right).$$

The values $\tilde{\mu}_i = y_i$ produce the highest likelihood. Thus, the deviance function is

$$D(y; \nu, \hat{\mu}) = 2\{\ell(\nu, \tilde{\mu}; y) - \ell(\nu, \hat{\mu}; y)\}$$

$$= 2\nu\sum_{i=1}^{n}\left(\ln\left(\frac{\hat{\mu}_i}{y_i}\right) - \frac{(y_i - \hat{\mu}_i)}{\hat{\mu}_i}\right). \tag{9.26}$$

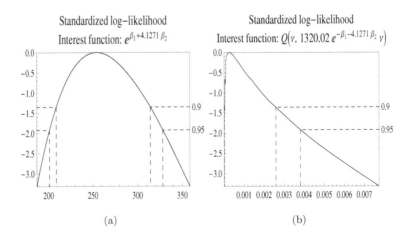

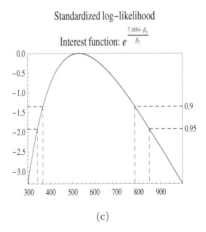

FIGURE 9.9: Brain weights data: Plots of log-likelihoods of interest functions (a) ψ_1, (b) ψ_2, and (c) ψ_3.

The asymptotic distribution of the deviance $D(y; \nu, \hat{\mu})$ is χ^2-distribution with $n - d$ degrees of freedom.

Example 9.8 Brain weights data (cont'd)

Let us test the statistical hypothesis H : $\nu = 1$, that is, the assumption that brain weights would have exponential distribution. The log-likelihood of the full and restricted models are -268.752 and -275.408, respectively. So the deviance 13.313 with 1 degree of freedom gives 0.000264 as the observed significance level.   ⬜

9.2.5   Model checking

Assume that the mean of ith component of the response vector is some nonlinear function of the regression parameters

$$\mu_i = \eta_i = \eta_i(\beta).$$

Then the deviance residual of the ith component of the response vector is

$$d_i = \text{sign}(y_i - \hat{\mu}_i) \left[2\nu \left\{ \ln \left(\frac{\hat{\mu}_i}{y_i} \right) - \frac{(y_i - \hat{\mu}_i)}{\hat{\mu}_i} \right\} \right]^{1/2}, \tag{9.27}$$

where

$$\hat{\mu}_i = \eta_i(\hat{\beta}).$$

The hat matrix is equal to

$$H(\hat{\beta}) = W(\hat{\beta})^{1/2} X(\hat{\beta}) (X(\hat{\beta})^{\text{T}} W(\hat{\beta}) X(\hat{\beta}))^{-1} X(\hat{\beta})^{\text{T}} W(\hat{\beta})^{1/2},$$

where

$$X(\beta) = \frac{\eta(\beta)}{\partial \beta^{\text{T}}} = \left(\frac{\eta_i(\beta)}{\partial \beta_j} \right)$$

and

$$W(\beta) = \text{diag} \left(\frac{1}{\eta_1(\beta)^2}, \ldots, \frac{1}{\eta_n(\beta)^2} \right). \tag{9.28}$$

The standardized deviance residuals have the form

$$r_{Di} = \frac{d_i}{\sqrt{1 - h_{ii}}}. \tag{9.29}$$

The standardized Pearson residual is

$$r_{Pi} = \frac{u_i(\hat{\beta})}{\sqrt{w_i(\hat{\beta})(1 - h_{ii})}} = \frac{\nu(y_i - \hat{\mu}_i)}{\hat{\mu}_i \sqrt{1 - h_{ii}}}. \tag{9.30}$$

Detailed calculations show that the distributions of the quantities

$$r_i^* = r_{Di} + \frac{1}{r_{Di}} \ln \left(\frac{r_{Pi}}{r_{Di}} \right) \tag{9.31}$$

are close to normal.

Example 9.9 Brain weights data (cont'd)

Consider the gamma regression model with the log-link function and with the linear effect of logarithm of body weight. Figure 9.10 gives plots of standardized deviance, Pearson, and modified residuals. Figure 9.11 contains plots of modified residuals against species index, logarithm of body weight, and fitted value and also observed brain weights and estimated mean curve of brain weight against the logarithm of body weight. Two things stick out in these plots. First, in all the humans, that is, "*Homo sapiens*," seems to be an "outlier." Second, either the model is not working well in the case of large body weights or the African elephant may be another "outlier." ⬜

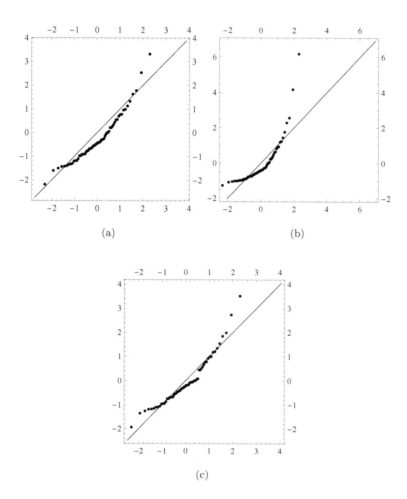

FIGURE 9.10: Brain weights data: Plots of residuals: (a) standardized deviance residuals, (b) standardized Pearson residuals, and (c) modified residuals.

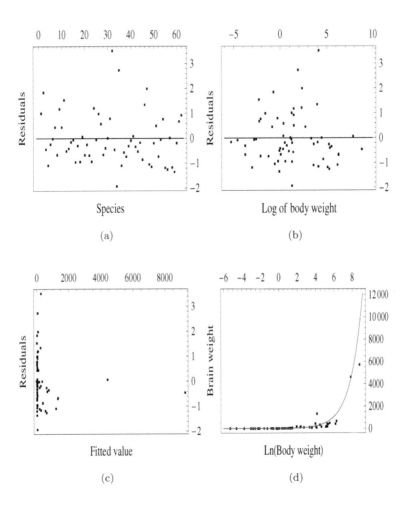

FIGURE 9.11: Brain weights data: Plots of modified residuals and mean brain weight: (a) modified residuals against species index, (b) modified residuals against logarithm of body weight, (c) modified residuals against fitted values, and (d) observed and mean brain weight against the logarithm of body weight.

Chapter 10

Other Generalized Regression Models

10.1 Weighted general linear model

10.1.1 Model

The linear regression model

$$y_i = x_i\beta + \varepsilon_i \ \forall i = 1, \ldots, n \tag{10.1}$$

or, in matrix form,

$$y = X\beta + \varepsilon \tag{10.2}$$

is a *weighted general linear model* if

$$\mathrm{E}(y) = X\beta \tag{10.3}$$

and

$$\mathrm{var}(y) = \sigma^2 \mathrm{diag}\left(\frac{1}{w_1}, \frac{1}{w_2}, \ldots, \frac{1}{w_n}\right), \tag{10.4}$$

where real numbers $w_1, w_2, \ldots, w_n$ are known *weights*. If, in addition, the components of the response vector are assumed to be statistically independent and normally distributed, then the likelihood function of the model is

$$L(\beta, \sigma; y) = \prod_{i=1}^{n} \frac{\sqrt{w_i}}{\sqrt{2\pi}\sigma} e^{-\frac{w_i(y_i - x_i\beta)^2}{2\sigma^2}} \tag{10.5}$$

and the log-likelihood function

$$\ell(\beta, \sigma; y) = -\frac{n}{2}\ln(2\pi) - n\ln(\sigma) + \frac{1}{2}\sum_{i=1}^{n}\ln(w_i) - \frac{1}{2\sigma^2}\sum_{i=1}^{n}w_i(y_i - x_i\beta)^2$$

$$= -n\ln(\sigma) - \frac{1}{2\sigma^2}(y - X\beta)^{\mathrm{T}}W(y - X\beta), \tag{10.6}$$

where W is the weight matrix

$$W = \mathrm{diag}(w_1, w_2, \ldots, w_n).$$

As the weighted linear regression model is a special case of what was considered in Chapter 3, all theoretical results apply to it. So, for example, the least squares (maximum likelihood) estimate of the regression coefficients has the form

$$\hat{\beta} = (X^{\mathrm{T}}W^{-1}X)^{-1}X^{\mathrm{T}}W^{-1}y \tag{10.7}$$

and its mean vector is

$$E(\hat{\beta}) = \beta, \tag{10.8}$$

variance matrix

$$\mathrm{var}(\hat{\beta}) = \sigma^2(X^{\mathrm{T}}W^{-1}X)^{-1}, \tag{10.9}$$

and if the response y is normally distributed it has the following normal distribution:

$$\hat{\beta} = (X^{\mathrm{T}}W^{-1}X)^{-1}X^{\mathrm{T}}W^{-1}y$$
$$\sim \mathrm{MultivariateNormalModel}[d, \beta, \sigma^2(X^{\mathrm{T}}W^{-1}X)^{-1}].$$

10.1.2 Weighted linear regression model as GRM

The weighted linear regression model is a two-parameter GRM such that

$$\mu_i = x_i\beta = x_{i1}\beta_1 + \ldots + x_{id}\beta_d \tag{10.10}$$

with $n \times d$ matrix X as the model matrix for the means and

$$\sigma_i = z_i\sigma = \frac{1}{\sqrt{w_i}}\sigma \tag{10.11}$$

with $n \times 1$ matrix

$$Z = \left(\frac{1}{\sqrt{w_1}}, \frac{1}{\sqrt{w_2}}, \ldots, \frac{1}{\sqrt{w_n}}\right)^{\mathrm{T}}$$

as the model matrix for the standard deviations.

Example 10.1 Fuel consumption data (cont'd)

Let us assume that observations of fuel consumption are statistically independent normally distributed so that their means depend linearly on air temperature and wind velocity and thus the model matrix for the means is

$$X = \begin{pmatrix} 1 & -3.0 & 15.3 \\ 1 & -1.8 & 16.4 \\ 1 & -10.0 & 41.2 \\ \vdots & \vdots & \vdots \\ 1 & -4.2 & 24.3 \\ 1 & -8.8 & 14.7 \\ 1 & -2.3 & 16.1 \end{pmatrix}.$$

TABLE 10.1: Fuel consumption data: Estimates and approximate 0.95-level confidence intervals of model parameters.

Parameter	Estimate	Lower limit	Upper limit
β_1	12.67	10.46	14.90
β_2	-0.591	-0.762	-0.419
β_3	0.114	0.0159	0.2108
σ	2.613	1.787	4.375

Let us also assume that the standard deviations of the observations were not constant but instead had weights $1, 2, \ldots$, and 10. Thus, the model matrix for the standard deviations is

$$
Z = \begin{pmatrix} 1 \\ \frac{1}{\sqrt{2}} \\ \frac{1}{\sqrt{3}} \\ \vdots \\ \frac{1}{2\sqrt{2}} \\ \frac{1}{3} \\ \frac{1}{\sqrt{10}} \end{pmatrix}.
$$

The statistical model for the observations of fuel consumption is defined by

$$
\mathcal{M} = \text{RegressionModel}[\{X, Z\}].
$$

The observed log-likelihood function of the model $\mathcal{M}$ is

$$
-\frac{1}{\sigma^2}\left(27.5\beta_1^2 - 341.4\beta_2\beta_1 + 904.4\beta_3\beta_1 - 1001.41\beta_1 - 5228.02\beta_2\beta_3 - \right.
$$
$$
16637.5\beta_3 + 1629.18\beta_2^2 + 9193.7\beta_3^2 + 6844.85\beta_2 + 9346.17\left.\right) -
$$
$$
10.\ln(\sigma) - 1.63718
$$

Table 10.1 gives MLEs and approximate 0.95-level profile likelihood-based confidence intervals for the regression coefficients and standard deviation. Figure 10.1 gives plots of log-likelihoods of model parameters. Exact Student t-intervals for all the regression coefficients are (9.78,15.6), (-0.814,-0.367), and (-0.0139,0.241), respectively. F-test for the statistical hypothesis of no linear wind velocity effect gives an observed significance level of 0.0732. The test of the statistical hypothesis of no air temperature and wind velocity effect, that is, the constant mean model, gives an observed significance level of 0.00129.

☐

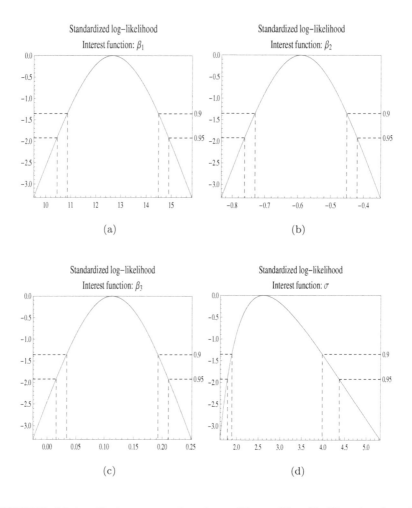

FIGURE 10.1: Fuel consumption data: Plots of log-likelihoods of model parameters (a) β_1, (b) β_2, (c) β_3, and (d) σ.

10.2   Weighted nonlinear regression model

10.2.1   Model

The nonlinear regression model

$$y_i = f(x_i; \beta) + \varepsilon_i \ \forall i = 1, \ldots, n \tag{10.12}$$

or, in matrix form,

$$y = f(X; \beta) + \varepsilon \tag{10.13}$$

is a weighted nonlinear regression model if

$$E(y) = f(X; \beta) \tag{10.14}$$

and

$$\mathrm{var}(y) = \sigma^2 \mathrm{diag}\left(\frac{1}{w_1}, \frac{1}{w_2}, \ldots, \frac{1}{w_n}\right), \tag{10.15}$$

where real numbers $w_1, w_2, \ldots, w_n$ are known *weights*. If, in addition, the components of the response vector are assumed to be statistically independent and normally distributed, then the likelihood function of the model is

$$L(\beta, \sigma; y) = \prod_{i=1}^{n} \frac{\sqrt{w_i}}{\sqrt{2\pi}\sigma} e^{-\frac{w_i(y_i - f(x_i; \beta))^2}{2\sigma^2}} \tag{10.16}$$

and the log-likelihood function

$$\ell(\beta, \sigma; y) = -\frac{n}{2}\ln(2\pi) - n\ln(\sigma) + \frac{1}{2}\sum_{i=1}^{n}\ln(w_i) - \frac{1}{2\sigma^2}\sum_{i=1}^{n}w_i(y_i - f(x_i; \beta))^2$$

$$= -\frac{n}{2}\ln(2\pi) - n\ln(\sigma) + \frac{1}{2}\sum_{i=1}^{n}\ln(w_i) -$$

$$\frac{1}{2\sigma^2}(y - f(X; \beta))^{\mathrm{T}}W(y - f(X; \beta)), \tag{10.17}$$

where W is the weight matrix

$$W = \mathrm{diag}(w_1, w_2, \ldots, w_n).$$

As a weighted nonlinear regression model is a special case of what was considered in Chapter 4, all theoretical results apply to it.

10.2.2   Weighted nonlinear regression model as GRM

A weighted nonlinear regression model is a two-parameter GRM such that

$$\mu_i = f(x_i; \beta) \tag{10.18}$$

and

$$\sigma_i = \frac{1}{\sqrt{w_i}} \sigma. \tag{10.19}$$

Example 10.2 Puromycin data (cont'd)

Consider the treated group and assume that observations have weights (1.1, 3.4, 2.8, 1.7, 3.2, 1.2, 3.6, 2.5, 1.4, 1.5, 2.9, 2.5). The model matrix for the standard deviations is

$$Z = \begin{pmatrix} 0.953463 \\ 0.542326 \\ 0.597614 \\ \vdots \\ 0.816497 \\ 0.587220 \\ 0.632456 \end{pmatrix}$$

and their regression function is

$$g(z; \sigma) = z\sigma.$$

The nonlinear weighted regression model is denoted by

$$\mathcal{M} = \text{RegressionModel}\left[\{X, Z\}, \left\{\frac{\beta_1 x}{\beta_2 + x}, z\sigma\right\}, \{\{x\}, \{z\}\}, \{\{\beta_1, \beta_2\}, \{\sigma\}\}\right].$$

The log-likelihood function is

$$-\frac{1}{\sigma^2}\left(0.55\left(y_1 - \frac{0.02\beta_1}{\beta_2 + 0.02}\right)^2 + 1.7\left(y_2 - \frac{0.02\beta_1}{\beta_2 + 0.02}\right)^2 + \right.$$

$$1.4\left(y_3 - \frac{0.06\beta_1}{\beta_2 + 0.06}\right)^2 + \ldots + 0.75\left(y_{10} - \frac{0.56\beta_1}{\beta_2 + 0.56}\right)^2 +$$

$$\left. 1.45\left(y_{11} - \frac{1.1\beta_1}{\beta_2 + 1.1}\right)^2 + 1.25\left(y_{12} - \frac{1.1\beta_1}{\beta_2 + 1.1}\right)^2\right) - 12\ln(\sigma) - 6.45478.$$

Assume that in addition to model parameters, the mean of velocity at 0.3 and concentration at which the mean is equal to 100 are also of interest, that is, parameter functions

$$\psi_1 = \frac{0.3\beta_1}{\beta_2 + 0.3}$$

TABLE 10.2: Puromycin data: Estimates and approximate 0.95-level confidence intervals of interest functions in the treated group.

Parameter	Estimate	Lower limit	Upper limit
β_1	214.4	203.1	226.9
β_2	0.0706	0.0570	0.0875
σ	12.52	8.77	19.81
ψ_1	173.5	167.7	179.5
ψ_2	0.0618	0.0532	0.0714

and

$$\psi_2 = \frac{100\beta_2}{\beta_1 - 100}.$$

Table 10.2 gives MLEs and approximate 0.95-level profile likelihood-based confidence intervals for model parameters and interest functions. For regression parameters confidence intervals with better coverage should be those that are based on appropriate F-distribution and these intervals are (201.5, 228.7) and (0.0551, 0.0902), respectively. Figure 10.2 gives the log-likelihood functions of interest functions. ⬚

10.3   Quality design or Taguchi model

The linear regression model

$$y_i = x_i\beta + \varepsilon_i \ \forall i = 1, \ldots, n \tag{10.20}$$

or, in matrix form,

$$y = X\beta + \varepsilon \tag{10.21}$$

is a quality design or Taguchi model if in addition to the linear model for the mean vector

$$\mathrm{E}(y) = \mu = X\beta \tag{10.22}$$

the transformed standard deviations of the responses are assumed to depend linearly on some explanatory factors, that is, for example

$$\ln(\sigma_i) = z_{i1}\gamma_1 + z_{i2}\gamma_2 + \ldots + z_{iq}\gamma_q = z_i\gamma \tag{10.23}$$

or

$$\mathrm{var}(y) = \mathrm{diag}(\sigma_1^2, \sigma_2^2, \ldots, \sigma_n^2) = \mathrm{diag}\left(e^{2z_1\gamma}, e^{2z_2\gamma}, \ldots, e^{2z_n\gamma}\right). \tag{10.24}$$

The $n \times q$ model matrix for the standard deviation is denoted by Z and $q \times 1$ vector γ contains the unknown regression coefficients of the standard deviations.

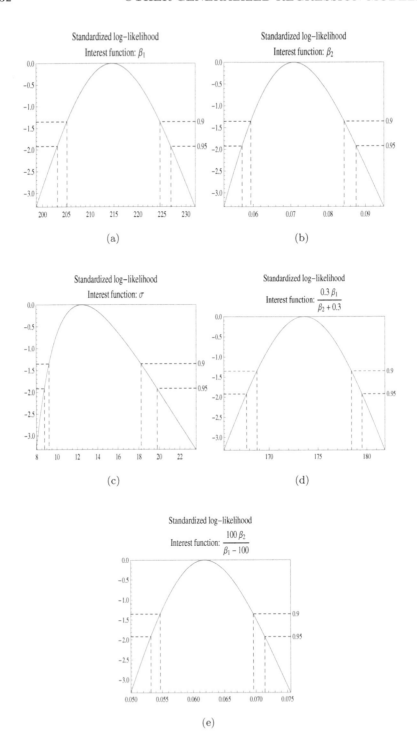

FIGURE 10.2:   Puromycin data: Plots of log-likelihoods of interest functions (a) β_1, (b) β_2, (c) σ, (d) ψ_1, and (e) ψ_2 in the treated group.

If, in addition, the components of the response vector are assumed to be statistically independent and normally distributed, then the likelihood function of the model is

$$L(\beta, \gamma; y) = \prod_{i=1}^{n} \frac{1}{\sqrt{2\pi e^{z_i \gamma}}} e^{-\frac{(y_i - x_i \beta)^2}{2 e^{2 z_i \gamma}}} \tag{10.25}$$

and the log-likelihood function

$$\ell(\beta, \gamma; y) = -\frac{n}{2} \ln(2\pi) - \sum_{i=1}^{n} z_i \gamma - \frac{1}{2} \sum_{i=1}^{n} \frac{(y_i - x_i \beta)^2}{e^{2 z_i \gamma}}$$

$$= -\frac{n}{2} \ln(2\pi) - \sum_{i=1}^{n} z_i \gamma - \frac{1}{2} (y - X\beta)^{\mathrm{T}} V (y - X\beta), \tag{10.26}$$

where V is the matrix

$$V = \mathrm{diag}\left(\frac{1}{e^{2 z_1 \gamma}}, \frac{1}{e^{2 z_2 \gamma}}, \ldots, \frac{1}{e^{2 z_n \gamma}} \right). \tag{10.27}$$

Note that the matrix V contains unknown parameters and so the theory of general linear models does not apply to these models, but instead in statistical inference one must rely on the general likelihood-based approach considered in Chapter 1.

The Taguchi model is a two-parameter GRM such that

$$\mu_i = x_i \beta = x_{i1} \beta_1 + \ldots + x_{id} \beta_d \tag{10.28}$$

with $n \times d$ matrix X as the model matrix for the means and

$$\sigma_i = e^{z_{i1} \gamma_1 + z_{i2} \gamma_2 + \ldots + z_{iq} \gamma_q} = e^{z_i \gamma} \tag{10.29}$$

with $n \times 1$ matrix Z as the model matrix for the standard deviations.

Example 10.3 Leaf springs data (cont'd)

The values of height measurements form the response and the vectors for the effects of oil temperature (O), transfer time (D), heating time (C), furnace temperature (B), and "hold-down" time (E) are presented in Table 10.3.

Because Figure 2.1 on page 54 in Chapter 2 indicated that oil temperature, transfer time, and furnace temperature affect the mean, the model matrix for the mean is defined as shown in Table 10.4 and Figure 2.2 on page 55 that heating time, furnace temperature, and hold-down time affect the standard deviation, the model matrix for the standard deviation is defined as shown in Table 10.5.

The Taguchi model is denoted by

$$\mathcal{M} = \mathrm{RegressionModel}[\{X, Z\},$$
$$\mathrm{Distribution} \rightarrow \mathrm{SamplingModel}[\mathrm{NormalModel}[\mu, \sigma], 3],$$
$$\mathrm{InverseLink} \rightarrow \{z, e^z\}].$$

TABLE 10.3: Leaf springs data: Levels of factors.

Unit	O	D	C	B	E
1	-1	-1	-1	-1	-1
2	-1	-1	-1	1	1
3	-1	-1	1	-1	1
4	-1	-1	1	1	-1
5	-1	1	-1	-1	1
6	-1	1	-1	1	-1
7	-1	1	1	-1	-1
8	-1	1	1	1	1
9	1	-1	-1	-1	-1
10	1	-1	-1	1	1
11	1	-1	1	-1	1
12	1	-1	1	1	-1
13	1	1	-1	-1	1
14	1	1	-1	1	-1
15	1	1	1	-1	-1
16	1	1	1	1	1

TABLE 10.4: Leaf springs data: Model matrix for means.

Unit	Constant	O	D	B	O.D	O.B	O.D.B
1	1	-1	-1	-1	1	1	-1
2	1	-1	-1	1	1	-1	1
3	1	-1	-1	-1	1	1	-1
4	1	-1	-1	1	1	-1	1
5	1	-1	1	-1	-1	1	1
6	1	-1	1	1	-1	-1	-1
7	1	-1	1	-1	-1	1	1
8	1	-1	1	1	-1	-1	-1
9	1	1	-1	-1	-1	-1	1
10	1	1	-1	1	-1	1	-1
11	1	1	-1	-1	-1	-1	1
12	1	1	-1	1	-1	1	-1
13	1	1	1	-1	1	-1	-1
14	1	1	1	1	1	1	1
15	1	1	1	-1	1	-1	-1
16	1	1	1	1	1	1	1

TABLE 10.5: Leaf springs data: Model matrix for standard deviations.

Unit	Constant	C	B	E	C.B	B.e	C.B.E
1	1	-1	-1	-1	1	1	-1
2	1	-1	1	1	-1	1	-1
3	1	1	-1	1	-1	-1	-1
4	1	1	1	-1	1	-1	-1
5	1	-1	-1	1	1	-1	1
6	1	-1	1	-1	-1	-1	1
7	1	1	-1	-1	-1	1	1
8	1	1	1	1	1	1	1
9	1	-1	-1	-1	1	1	-1
10	1	-1	1	1	-1	1	-1
11	1	1	-1	1	-1	-1	-1
12	1	1	1	-1	1	-1	-1
13	1	-1	-1	1	1	-1	1
14	1	-1	1	-1	-1	-1	1
15	1	1	-1	-1	-1	1	1
16	1	1	1	1	1	1	1

Table 10.6 contains MLEs and approximate 0.95-level profile likelihood-based confidence intervals of the model parameters. The observed significance level of the likelihood ratio test of the statistical hypothesis $H : \gamma_C = 0, \gamma_{C.B} = 0, \gamma_{C.B.E} = 0$ is 0.197. Table 10.7 gives estimates and approximate 0.95-level profile likelihood-based confidence intervals for the submodel. These estimates indicate that the variation of heights of leaf springs can first be minimized using heating time and furnace temperature and then keeping values of these factors fixed adjust the mean height to some target value with the help of oil temperature and transfer time because these factors do not affect the variation. Table 10.8 contains estimates and confidence intervals for standard deviations of various combinations of heating time and furnace temperature appearing in the data. Heating time at low level and furnace temperature at high level minimizes variation. Similarly, Table 10.9 shows estimates and confidence intervals for means of various combinations of oil temperature, transfer time, and furnace temperature appearing in the data. The target value 8 in. can be nearly achieved by keeping furnace temperature fixed at high levels and selecting for oil temperature the low level and for transfer time the high level. □

TABLE 10.6:   Leaf springs data: Estimates and approximate 0.95-level confidence intervals of model parameters.

Parameter	Estimate	Lower limit	Upper limit
$\beta_{Constant}$	7.648	7.6136	7.681
β_O	-0.0954	-0.131	-0.0569
β_D	0.0663	0.0342	0.0891
β_B	0.0594	0.0284	0.0910
$\beta_{O.D}$	-0.0930	-0.128	-0.0542
$\beta_{O.B}$	0.0393	0.00355	0.0777
$\beta_{O.D.B}$	0.0390	0.00381	0.0776
$\gamma_{Constant}$	-2.004	-2.191	-1.789
γ_C	-0.221	-0.472	0.00943
γ_B	0.221	0.0104	0.429
γ_E	-0.282	-0.509	-0.0595
$\gamma_{C.B}$	-0.0863	-0.321	0.144
$\gamma_{B.E}$	0.710	0.459	0.937
$\gamma_{C.B.E}$	0.0856	-0.126	0.295

TABLE 10.7:   Leaf springs data: Estimates and approximate 0.95-level confidence intervals of model parameters in submodel.

Parameter	Estimate	Lower limit	Upper limit
$\beta_{Constant}$	7.657	7.624	7.692
β_O	-0.108	-0.143	-0.0750
β_D	0.0678	0.0405	0.0908
β_B	0.0650	0.0319	0.100
$\beta_{O.D}$	-0.0969	-0.130	-0.0576
$\beta_{O.B}$	0.0282	-0.00594	0.0617
$\beta_{O.D.B}$	0.0343	0.000681	0.0734
$\gamma_{Constant}$	-1.955	-2.14243	-1.741
γ_B	0.232	0.0279	0.436
γ_E	-0.248	-0.466	-0.0336
$\gamma_{B.E}$	0.703	0.471	0.918

TABLE 10.8:   Leaf springs data: Estimates and approximate 0.95-level confidence intervals of standard deviations for combinations of heating time and furnace temperature factors.

Combination	Estimate	Lower limit	Upper limit
(-1,-1)	0.291	0.202	0.464
(-1, 1)	0.0433	0.0301	0.0708
(1,-1)	0.113	0.0787	0.185
(1, 1)	0.282	0.192	0.454

TABLE 10.9: Leaf springs data: Estimates and approximate 0.95-level confidence intervals of means for combinations of oil temperature, transfer time, and furnace temperature factors.

Combination	Estimate	Lower limit	Upper limit
(-1,-1,-1)	7.530	7.480	7.586
(-1,-1,1)	7.672	7.561	7.802
(-1,1,-1)	7.928	7.869	7.977
(-1,1,1)	7.933	7.817	8.043
(1,-1,-1)	7.519	7.465	7.570
(1,-1,1)	7.636	7.520	7.746
(1,1,-1)	7.392	7.337	7.442
(1,1,1)	7.647	7.538	7.771

10.4 Lifetime regression model

Let

$$z_i \sim f(z_i; \omega_i) \ \forall i = 1, \ldots, n \qquad (10.30)$$

be statistically independent quantities and assume that observations consist of pairs (y_i, c_i) of numbers such that c_i has two possible values 0 and 1. If quantity c_i takes value 0, then $y_i = z_i$ meaning that quantity z_i has been observed as such. However, when c_i takes value 1 the quantity z_i has been censored and it is only known that $z_i > y_i$. Statistical variable c is called the *censoring status* variable. Usually censoring is such that in censored cases y_i gives a lower bound to z_i, but also other types of censoring occur, for example, y_i might be an upper bound or y_i might even be an interval consisting of a lower and an upper bound. In fact, one might say that because of finite measurement accuracy exact observations never occur and so all observations are more or less censored with the latter type censoring being the most common. The following discussion applies to any type of censoring, but for the sake of simplicity in formulas lower bound censoring is assumed. The likelihood function has the form

$$L(\omega; y) = \prod_{i=1}^{n} f(y_i; \omega_i)^{1-c_i} (1 - F(y_i; \omega_i))^{c_i}, \qquad (10.31)$$

where F denotes the cumulative distribution function of z. Now, if the parameters ω_i have the form

$$\omega_i = (\eta(x_i; \beta), \phi)$$

TABLE 10.10:   Carcinogen data.

Group	Number of days
1	143, 164, 188, 188, 190, 192, 206, 209, 213, 216, 220, 227, 230, 234, 246, 265, 304, 216+, 244+
2	142, 156, 163, 198, 205, 232, 232, 233, 233, 233, 233, 239, 240, 264, 280, 280, 296, 296, 323, 204+, 344+

for some known vectors x_i of real numbers, unknown vectors β and ϕ, and known function η, the log-likelihood function of the statistical evidence is

$$\ell(\beta, \phi; y) = \sum_{i=1}^{n} [(1 - c_i) \ln(f(y_i; \eta(x_i; \beta), \phi)) + c_i \ln(1 - F(y_i; \eta(x_i; \beta), \phi))].$$

$$(10.32)$$

Example 10.4 Carcinogen data

Two groups of rats were exposed to the carcinogen DBMA, and the number of days to death due to cancer was recorded (Kalbfleisch and Prentice 1980). The columns of Table 10.10 contain values of group and number of days. Censored times have been denoted by a + sign. Figure 10.3 shows values of number of days.

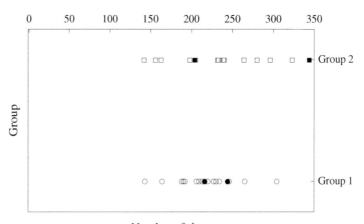

FIGURE 10.3:   Carcinogen data: Plot of the data.

TABLE 10.11:   Carcinogen data: Estimates and
approximate 0.95-level confidence intervals of model
parameters.

Parameter	Estimate	Lower limit	Upper limit
β_1	5.469	5.376	5.564
β_2	-0.0936	-0.230	0.0429
σ	0.210	0.169	0.271

The model matrix has the following simple form

$$X = \begin{pmatrix} 1 & 1 \\ 1 & 1 \\ 1 & 1 \\ \vdots & \vdots \\ 1 & 0 \\ 1 & 0 \\ 1 & 0 \end{pmatrix}.$$

and assuming log-normal distribution for z_is the statistical model is defined
by

$$\mathcal{M} = \text{RegressionModel}[X, \text{Distribution} \rightarrow \text{LogNormalModel}[\mu, \sigma]].$$

The log-likelihood function is

$$-\frac{(\ln(143) - \beta_1 - \beta_2)^2}{2\sigma^2} - \frac{(\ln(164) - \beta_1 - \beta_2)^2}{2\sigma^2}$$
$$-\frac{(\ln(188) - \beta_1 - \beta_2)^2}{2\sigma^2} - \cdots - \frac{(\ln(323) - \beta_1)^2}{2\sigma^2}$$
$$+ \ln\left(1 - \Phi\left(\frac{\ln(204) - \beta_1}{\sigma}\right)\right) + \ln\left(1 - \Phi\left(\frac{\ln(344) - \beta_1}{\sigma}\right)\right) - 36\ln(\sigma).$$

Table 10.11 gives MLEs and approximate 0.95-level profile likelihood-based
confidence intervals of the regression and shape parameters. The observed
significance test of the statistical hypothesis of no group effect is 0.173.

Consider next the model without group effect, that is, assume observations to be a censored sample from some log-normal distribution. In addition
to model parameters assume that of interest are mean, standard deviation,
median, and interquartile deviance scaled to correspond to that of normal
distribution, that is, following parameter functions

$$\psi_1 = e^{\mu + \frac{\sigma^2}{2}},$$
$$\psi_2 = e^{\mu + \frac{\sigma^2}{2}} \sqrt{e^{\sigma^2} - 1},$$
$$\psi_3 = e^{\mu},$$
$$\psi_4 = 0.741301(e^{\mu + 0.67449\sigma} - e^{\mu - 0.67449\sigma}).$$

TABLE 10.12: Carcinogen data: Estimates and approximate 0.95-level confidence intervals of interest functions in the submodel.

Parameter function	Estimate	Lower limit	Upper limit
μ	5.425	5.356	5.496
σ	0.216	0.174	0.278
ψ_1	232.4	217.1	250.6
ψ_2	50.79	40.01	68.01
ψ_3	227.0	211.9	243.7
ψ_4	49.21	39.19	64.62

Table 10.12 gives MLEs and approximate 0.95-level profile likelihood-based confidence intervals for model parameters and interest functions. Figure 10.4 shows plots of log-likelihood functions of model parameters and interest functions. ⬚

10.5   Cox regression model

Let $(y_i, c_i) \; \forall i = 1, \ldots, n$ be as above with the assumption that in censored cases y_i is a lower bound. Assume also that regressors affect the survivor function in the following way:

$$S(z; x) = 1 - F(z; x) = S_0(z)^{\eta(x;\beta)} \tag{10.33}$$

where F is the cumulative distribution function, S_0 is the so-called base line survivor function and, η is a known function of factors and regression parameters taking non-negative values. Hazard rate at z is

$$h(z; x) = -\frac{d}{dz} \ln(S(z; x)) = \eta(x; \beta)h_0(z), \tag{10.34}$$

where h_0 is the base line hazard function. The statistical model is the so-called *proportional hazard regression model* or *Cox regression model* because for any two values x and $\tilde{x}$ the ratio of hazards is equal to

$$\frac{h(z; x)}{h(z; \tilde{x})} = \frac{\eta(x; \beta)h_0(z)}{\eta(\tilde{x}; \beta)h_0(z)} = \frac{\eta(x; \beta)}{\eta(\tilde{x}; \beta)}$$

which is independent of z. A common choice for η is

$$\eta(x; \beta) = e^{x^{\mathsf{T}}\beta}.$$

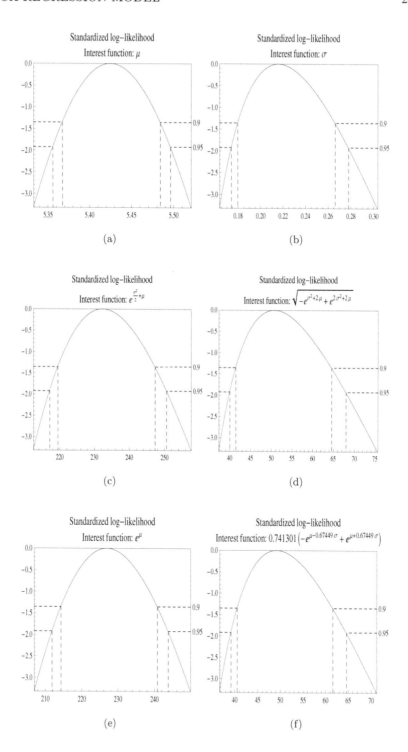

FIGURE 10.4: Carcinogen data: Plots of log-likelihoods of interest functions (a) μ, (b) σ, (c) ψ_1, (d) ψ_2, (e) ψ_3, and (f) ψ_4 in the submodel.

Denoting by z_x and $z_{\tilde{x}}$ statistically independent lifetimes with values x and $\tilde{x}$ of regressors we have

$$\text{pr}(z_x < z_{\tilde{x}}) = \frac{\eta(x;\beta)}{\eta(x;\beta) + \eta(\tilde{x};\beta)} \tag{10.35}$$

because

$$\text{pr}(z_x < z_{\tilde{x}}) = \int_0^\infty \text{pr}(t < z_{\tilde{x}}) f(t;x) dt$$

$$= \int_0^\infty S(t;\tilde{x}) h(t;x) S(t;x) dt$$

$$= \eta(x;\beta) \int_0^\infty h_0(t) S_0(t)^{\eta(\tilde{x};\beta)+\eta(x;\beta)} dt$$

and for any real number ζ and hazard function $h(t) = \zeta h_0(t)$ with survivor function $S(t) = S_0(t)^\zeta$

$$\int_0^\infty h_0(t) S_0(t)^\zeta dt = \frac{1}{\zeta} \int_0^\infty \zeta h_0(t) S_0(t)^\zeta dt$$

$$= \frac{1}{\zeta} \int_0^\infty h(t) S(t) dt$$

$$= \frac{1}{\zeta} \int_0^\infty f(t) dt = \frac{1}{\zeta}.$$

Davison (2003, pp. 543–544) shows that under certain natural conditions inference on regression parameters can be made using the so-called *partial likelihood function*

$$\prod_{i=1}^n \left(\frac{\eta(x_i;\beta)}{\sum_{j \in \mathcal{R}_i} \eta(x_j;\beta)} \right)^{1-c_i} = \prod_{\{i:c_i=0\}} \frac{\eta(x_i;\beta)}{\sum_{j \in \mathcal{R}_i} \eta(x_j;\beta)} \tag{10.36}$$

where $\mathcal{R}_i$ is the *risk set* for every uncensored observation y_i, that is, the set of all uncensored or censored observations j for which $y_j \geq y_i$. The derivation of this result is based on the profile likelihood method. One can also show that the partial likelihood function has the same probabilistic behavior as regular likelihood functions.

Statistical evidence which gives the same likelihood function consists of response vector $y_{\text{obs}} = (1, 1, \ldots, 1)$ arising from statistically independent Bernoulli trials corresponding to the uncensored observations y_i. Success probabilities of these Bernoulli distributions are equal to

$$\omega_i = \frac{\eta(x_i;\beta)}{\sum_{j \in \mathcal{R}_i} \eta(x_j;\beta)}.$$

TABLE 10.13:   Gastric cancer data.

Treatment	Time
CR	17, 42, 44, 48, 60, 72, 74, 95, 103, 108, 122, 144, 167, 170, 183, 185, 193, 195, 197, 208, 234, 235, 254, 307, 315, 401, 445, 464, 484, 528, 542, 567, 577, 580, 795, 855, 1174+, 1214, 1232+, 1366, 1455+, 1585+, 1622+, 1626+, 1736+
C	1, 63, 105, 125, 182, 216, 250, 262, 301, 301, 342, 354, 356, 358, 380, 383, 383, 388, 394, 408, 460, 489, 499, 523, 524, 535, 562, 569, 675, 676, 748, 778, 786, 797, 955, 968, 977, 1245, 1271, 1420, 1460+, 1516+, 1551, 1690+, 1694

Thus, the statistical model is defined by the expression

$$\mathcal{M} = \text{RegressionModel}\left[\left\{\frac{\eta(x_1; \beta)}{\sum_{j \in \mathcal{R}_1} \eta(x_j; \beta)}, \ldots, \frac{\eta(x_k; \beta)}{\sum_{j \in \mathcal{R}_k} \eta(x_j; \beta)}\right\}, \beta, \right.$$
$$\left. \text{Distribution} \rightarrow \text{BernoulliModel}[\omega]\right],$$

where it is assumed that uncensored observations have indexes 1, 2, ..., and k.

Example 10.5 Gastric cancer data

A total of 90 patients suffering from gastric cancer were randomly assigned to two groups. One group was treated with chemotherapy and radiation, whereas the other only received chemotherapy (Lindsey 1997, p. 118). The columns of Table 10.13 contain values of treatment and survival time in days. Censored times have been denoted by a + sign. Figure 10.5 contains a plot of the data.

Because data contains only one binary factor, the model matrix consists of one column containing the indicator of, for example, the group that was treated by both chemotherapy and radiation. Thus, the model matrix is

$$X = \begin{pmatrix} 1 \\ 1 \\ 1 \\ \vdots \\ 0 \\ 0 \\ 0 \end{pmatrix}.$$

FIGURE 10.5:   Gastric cancer data: Plot of the data.

The partial likelihood function has the form

$$\prod_{\{i:c_i=0\}} \frac{\eta(x_i;\beta)}{\sum_{j\in\mathcal{R}_i}\eta(x_j;\beta)} = \prod_{\{i:c_i=0\}} \frac{e^{\beta x_i}}{\sum_{j\in\mathcal{R}_i}e^{\beta x_j}}$$

where x_i is the indicator of the first group. For example, the three smallest uncensored observations are $y_{45} = 1$, $y_1 = 17$, and $y_2 = 42$, and the three largest, uncensored observations are $y_{85} = 1420$, $y_{88} = 1551$, and $y_{90} = 1694$, respectively. Their risk sets are

$$\mathcal{R}_{46} = \{1,\ldots,45,46,\ldots,90\},$$
$$\mathcal{R}_1 = \{1,\ldots,45,47,\ldots,90\},$$
$$\mathcal{R}_2 = \{2,\ldots,45,47,\ldots,90\},$$
$$\vdots$$
$$\mathcal{R}_{85} = \{41,42,43,44,45,85,86,87,88,89,90\},$$
$$\mathcal{R}_{88} = \{42,43,44,45,88,89,90\},$$
$$\mathcal{R}_{90} = \{90\}.$$

TABLE 10.14:   Gastric cancer data: Estimates and
approximate 0.95-level confidence intervals of interest functions.

Parameter function	Estimate	Lower limit	Upper limit
β	0.165	-0.279	0.605
ψ_1	1.179	0.757	1.832
ψ_2	0.541	0.431	0.647

Thus, the partial likelihood function has the form

$$\prod_{\{i:c_i=0\}} \frac{e^{\beta x_i}}{\sum_{j \in R_i} e^{\beta x_j}} =$$

$$\frac{e^{\beta 0}}{45e^{\beta 0} + 45e^{\beta 1}} \frac{e^{\beta 1}}{44e^{\beta 0} + 45e^{\beta 1}} \frac{e^{\beta 1}}{44e^{\beta 0} + 44e^{\beta 1}} \times \cdots \times$$

$$\frac{e^{\beta 0}}{6e^{\beta 0} + 5e^{\beta 1}} \frac{e^{\beta 0}}{3e^{\beta 0} + 4e^{\beta 1}} \frac{e^{\beta 0}}{1e^{\beta 0} + 1e^{\beta 1}} =$$

$$\frac{1}{45 + 45e^{\beta}} \frac{e^{\beta}}{44 + 45e^{\beta}} \frac{e^{\beta}}{44 + 44e^{\beta}} \times \cdots \times \frac{1}{6 + 5e^{\beta}} \frac{1}{3 + 4e^{\beta}} \frac{1}{1 + e^{\beta}}.$$

Consider in addition to parameter β interest functions

$$\psi_1 = e^{\beta}$$

and

$$\psi_2 = \frac{e^{\beta}}{1 + e^{\beta}}.$$

The first interest function is the hazard ratio of the group of those patients
who had both chemotherapy and radiation to the group of those patients
who had only chemotherapy. The second interest function is the probability
that two statistically independent observations from groups are such that an
observation from the first group is less than that from the second group. In
addition to the MLEs, Table 10.14 gives the approximate 0.95-level confidence
intervals for β and the two interest functions. In fact, in this case all three are
functions of each other and tell the same story that even though the estimate
of hazard ratio is greater than one, statistical evidence also supports hazard
ratios that are less than one. Figure 10.6 contains plots of log-likelihoods of
interest functions.   ☐

Example 10.6 Leukemia data (cont'd)

Because data contain only one binary factor, the model matrix consists of
one column containing an indicator of, for example, the drug group. Thus,

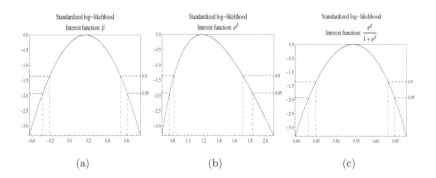

FIGURE 10.6: Gastric cancer data: Plots of log-likelihoods of interest functions (a) β, (b) ψ_1, and (c) ψ_2.

the model matrix is

$$
X = \begin{pmatrix} 1 \\ 1 \\ 1 \\ \vdots \\ 0 \\ 0 \\ 0 \end{pmatrix}.
$$

The partial likelihood function has the form

$$
\prod_{\{i:c_i=0\}} \frac{\eta(x_i;\beta)}{\sum_{j \in \mathcal{R}_i} \eta(x_j;\beta)} = \prod_{\{i:c_i=0\}} \frac{e^{\beta x_i}}{\sum_{j \in \mathcal{R}_i} e^{\beta x_j}}
$$

where x_i is an indicator of the drug group. For example, the three smallest uncensored observations are $y_{22}=1$, $y_{23}=1$, and $y_{24}=2$, and the three largest uncensored observations are $y_{15}=22$, $y_{16}=23$, and $y_{42}=23$, respectively. Their risk sets are

$$
\begin{aligned}
\mathcal{R}_{22} &= \{1,\ldots,21,22,\ldots,42\}, \\
\mathcal{R}_{23} &= \{1,\ldots,21,22,\ldots,42\}, \\
\mathcal{R}_{24} &= \{1,\ldots,21,24,\ldots,42\},
\end{aligned}
$$

$$\vdots$$

$$
\begin{aligned}
\mathcal{R}_{15} &= \{15,16,17,18,19,20,21,41,42\}, \\
\mathcal{R}_{16} &= \{16,17,18,19,20,21,42\}, \\
\mathcal{R}_{42} &= \{16,17,18,19,20,21,42\}.
\end{aligned}
$$

TABLE 10.15: Leukemia data: Estimates and approximate 0.95-level confidence intervals of interest functions.

Parameter function	Estimate	Lower limit	Upper limit
β	-1.509	-2.362	-0.737
ψ_1	0.221	0.0942	0.479
ψ_2	0.181	0.0861	0.324

Thus, the partial likelihood function has the form

$$\prod_{\{i:c_i=0\}} \frac{e^{\beta x_i}}{\sum_{j \in \mathcal{R}_i} e^{\beta x_j}} =$$

$$\frac{e^{\beta 0}}{21e^{\beta 0} + 21e^{\beta 1}} \frac{e^{\beta 0}}{21e^{\beta 0} + 21e^{\beta 1}} \frac{e^{\beta 0}}{19e^{\beta 0} + 21e^{\beta 1}} \times \ldots \times$$

$$\frac{e^{\beta 1}}{2e^{\beta 0} + 7e^{\beta 1}} \frac{e^{\beta 1}}{1e^{\beta 0} + 6e^{\beta 1}} \frac{e^{\beta 0}}{1e^{\beta 0} + 6e^{\beta 1}} =$$

$$\frac{1}{(21 + 21e^{\beta})^2} \frac{1}{(19 + 21e^{\beta})^2} \times \ldots \times \frac{e^{\beta}}{(2 + 7e^{\beta})^2} \frac{e^{\beta}}{(1 + 6e^{\beta})^2}.$$

Consider in addition to parameter β interest functions

$$\psi_1 = e^{\beta}$$

and

$$\psi_2 = \frac{e^{\beta}}{1 + e^{\beta}}.$$

The first interest function is the hazard ratio of the drug group to the control group. The second interest function is the probability that two statistically independent observations from groups are such that an observation from the first group is less than that from the second group. In addition to the MLEs, Table 10.15 gives approximate 0.95-level confidence intervals for β and the two interest functions. In fact, in this case all three are functions of each other and tell the same story that the drug group has a significantly lower hazard rate. For example, statistical evidence supports for ψ_2 values from 0.086 to 0.324, that is, ψ_2 might have a value as low as 0.086 and at most a value of 0.324. Figure 10.7 contains plots of log-likelihoods of interest functions. ☐

Example 10.7 Acute myelogeneous leukemia data

Survival times (weeks) were recorded for patients with acute myelogeneous leukemia, along with white blood cell (wbc) counts in thousands and AG-factors (Feigl and Zelen 1965; Lindsey 1997, p. 117). Patients were classified into two groups according to morphological characteristics of their wbc: AG positive had Auer rods and/or granulature of the leukemia cells in the bone

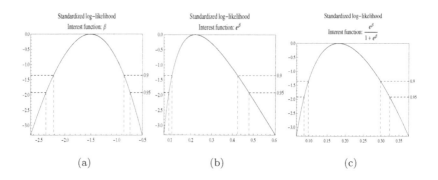

FIGURE 10.7:   Leukemia data: Plots of log-likelihoods of interest functions
(a) β, (b) ψ_1, and (c) ψ_2.

marrow at diagnosis, whereas AG negative did not. Table A.1 on page 252 in
Appendix A contains values of AG, wbc count, and survival time. Figure 10.8
shows a scatter plot of survival time against wbc count.

Because data contain one binary factor and one numeric factor, the model
matrix consists of one column containing an indicator of, for example, the AG
positive group and, assuming linear effect of wbc, two columns defining effect
of wbc for the AG positive group and difference of linear effects for groups,
respectively. Thus, the model matrix is

$$X = \begin{pmatrix} 1 & 2.3 & 2.3 \\ 1 & 10.5 & 10.5 \\ 1 & 9.4 & 9.4 \\ \vdots & & \\ 0 & 9.0 & 0 \\ 0 & 28.0 & 0 \\ 0 & 100.0 & 0 \end{pmatrix}.$$

Table 10.16 contains MLEs and approximate 0.95-level profile likelihood-based
confidence intervals of model parameters. Figure 10.9 shows plots of their log-
likelihood functions.   ⧠

10.6   Bibliographic notes

Box (1988) is an early article on quality design modeling.

Books on survival analysis include Kalbfleisch and Prentice (1980), Miller
(1981), Cox and Oakes (1984), Lee (1992), and Collet (1995).

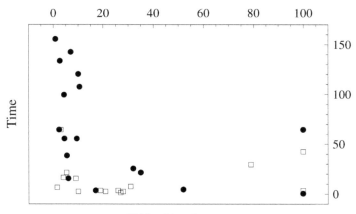

White blood count

FIGURE 10.8: Acute myelogeneous leukemia data: Scatter plot of data.

TABLE 10.16: Acute myelogeneous leukemia data: Estimates and approximate 0.95-level confidence intervals of model parameters.

Parameter	Estimate	Lower limit	Upper limit
β_1	-1.723	-2.804	-0.682
β_2	-0.00189	-0.0172	0.0112
β_3	0.0190	-0.000361	0.0388

Cox (1972) introduced proportional hazard modeling and Cox (1975) introduced the idea of partial likelihood.

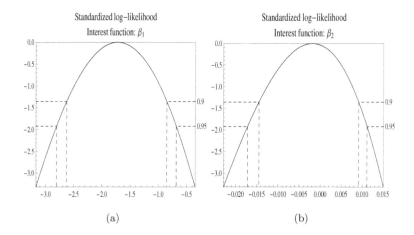

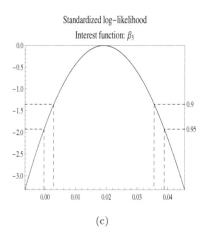

FIGURE 10.9: Acute myelogeneous leukemia data: Plots of log-likelihoods of model parameters (a) β_1, (b) β_2, and (c) β_3.

Appendix A

Datasets

Acute myelogeneous leukemia data

Survival times (weeks) were recorded for patients with acute myelogeneous leukemia, along with white blood cell (wbc) counts in the thousands and AG-factors (Feigl and Zelen 1965; Lindsey 1997, p. 117). Patients were classified into the two groups according to morphological characteristics of their wbc: AG positive had Auer rods and/or granulature of the leukemia cells in the bone marrow at diagnosis, whereas AG negative did not.

Table A.1 contains values of AG, wbc count, and survival time. Figure 10.8 on page 249 in Chapter 10 shows a scatter plot of survival time against wbc.

Bacterial concentrations data

These data arose from an investigation of various media to be used for storing microorganisms in a deep freeze. The results are bacterial concentrations (counts in a fixed area) measured at the initial freezing ($-70°C$) and then at 1, 2, 6, and 12 months afterwards (Francis *et al.* 1993, p. 531) and (Ludlam *et al.* 1989).

The columns in Table A.2 contain values of time and concentration. Figure 1.19 on page 39 in Chapter 1 plots a natural logarithm of count against time. How storing time affects the bacterial concentration was under investigation. The count seems to decrease somewhat with time.

Brain weights data

Table A.3 shows mean weights of brains and bodies of 62 species of animals. The columns contain values of species, brain weight in grams, and body weight in kilograms. Table A.4 gives explanations of species names.

Figure 9.5 on page 212 in Chapter 9 gives a graph of the natural logarithm of brain weight versus that of body weight showing a clear linear relation between these quantities.

Carcinogen data

Two groups of rats were exposed to the carcinogen DBMA, and the number of days to death due to cancer was recorded (Kalbfleisch and Prentice 1980).

TABLE A.1:   Acute
myelogeneous leukemia
data.

AG	Blood	Time
Positive	2.3	65
Positive	10.5	108
Positive	9.4	56
Positive	52.0	5
Positive	7.0	143
Positive	0.8	156
Positive	10.0	121
Positive	32.0	26
Positive	100.0	65
Positive	100.0	1
Positive	4.3	100
Positive	17.0	4
Positive	35.0	22
Positive	4.4	56
Positive	2.6	134
Positive	5.4	39
Positive	100.0	1
Positive	6.0	16
Negative	3.0	65
Negative	10.0	3
Negative	26.0	4
Negative	5.3	22
Negative	4.0	17
Negative	19.0	4
Negative	21.0	3
Negative	31.0	8
Negative	1.5	7
Negative	27.0	2
Negative	79.0	30
Negative	100.0	43
Negative	9.0	16
Negative	28.0	3
Negative	100.0	4

TABLE A.2:
Bacterial concentrations
data.

Time	Concentration
0	31
1	26
2	19
6	15
12	20

The columns of Table A.5 contain values of group and number of days to death. Censored times have been denoted by a + sign. Figure 10.3 on page 238 in Chapter 10 plots times of death.

Clotting time data

Hurn *et al.* (1945) published data on the clotting time of blood, giving clotting times in seconds for normal plasma diluted to nine different percentage concentrations with prothrombin-free plasma; clotting was induced by two lots of thromboplastin (McCullagh and Nelder 1989, p. 300).

The columns of Table A.6 contain values of lot, concentration, and time in seconds. Figure 2.4 on page 60 in Chapter 2 gives a plot of clotting times versus concentration for the two lots.

Coal miners data

Table A.7 shows time of exposure and prevalence of pneumoconiosis among a group of coal miners (Ashford 1959; McCullagh and Nelder 1989, p. 178).

The columns in Table A.7 contain values of time spent in a coal mine in years, and numbers of men in Category I (healthy), Category II (normal pneumoconiosis), and Category III (severe pneumoconiosis), respectively. Figure 2.7 on page 67 in Chapter 2 shows that cumulative odds for both pneumoconiosis and severe pneumoconiosis increase with time in a coal mine.

Fuel consumption data

The following data were collected to investigate how the amount of fuel oil required to heat a home depends upon the outdoor air temperature and wind velocity (Kalbfleisch 1985, p. 249).

The columns in Table A.8 contain values of fuel consumption, air temperature, and wind velocity. Figure 1.2 and Figure 1.3 on page 7 in Chapter 1 show scatter plots of variables. The effects of air temperature and wind

TABLE A.3: Brain weights data.

Species	Brain	Body	Species	Brain	Body
Ar. fox	44.500	3.385	Af. elephant	5711.860	6654.180
O. monkey	15.499	0.480	W. opossum	3.900	3.500
Beaver	8.100	1.350	R. monkey	179.003	6.800
Cow	423.012	464.983	Kangaroo	56.003	34.998
G. wolf	119.498	36.328	Y.-b. marmot	17.000	4.050
Goat	114.996	27.660	G. hamster	1.000	0.120
R. deer	98.199	14.831	Mouse	0.400	0.023
G. pig	5.500	1.040	L. b. bat	0.250	0.010
Vervet	57.998	4.190	S. loris	12.500	1.400
Chinchilla	6.400	0.425	Okapi	489.997	250.010
G. squirrel	4.000	0.101	Rabbit	12.100	2.500
Ar. g. squirrel	5.700	0.920	Sheep	175.002	55.501
Af. g. rat	6.600	1.000	Jaguar	156.993	100.003
S.-t. shrew	0.140	0.005	Chimpanzee	440.011	52.159
S.-n. mole	1.000	0.060	Baboon	179.504	10.550
9-b. armadillo	10.800	3.500	D. hedgehog	2.400	0.550
T. hyrax	12.300	2.000	G. armadillo	80.996	59.997
N. Am. opossum	6.300	1.700	R. hyrax2	21.000	3.600
As. elephant	4603.170	2547.070	Raccoon	39.201	4.288
B. b. bat	0.300	0.023	Rat	1.900	0.280
Donkey	419.012	187.092	S. Am. mole	1.200	0.075
Horse	654.977	521.026	M. rat	3.000	0.122
Eu. hedgehog	3.500	0.785	M. shrew	0.330	0.048
P. monkey	114.996	10.000	Pig	180.008	192.001
Cat	25.600	3.300	Echidna	25.001	3.000
Galago	5.000	0.200	B. tapir	169.000	160.004
Genet	17.500	1.410	Tenrec	2.600	0.900
G. seal	324.990	85.004	Phalanger	11.400	1.620
R. hyrax1	12.300	0.750	T. shrew	2.500	0.104
Human	1320.020	61.998	R. fox	50.400	4.235

TABLE A.4:   Species names.

Short name	Long name
9-b. armadillo	Nine-banded armadillo
Af. elephant	African elephant
Af. g. rat	African giant-pouched rat
Ar. fox	Arctic fox
Ar. g. squirrel	Arctic ground squirrel
As. elephant	Asian elephant
B. b. bat	Big brown bat
B. tapir	Brazilian tapir
D. hedgehog	Desert hedgehog
Eu. hedgehog	European hedgehog
G. armadillo	Giant armadillo
G. hamster	Golden hamster
G. pig	Guinea pig
G. seal	Gray seal
G. squirrel	Ground squirrel
G. wolf	Gray wolf
L. b. bat	Little brown bat
M. rat	Mole rat
M. shrew	Musk shrew
N. Am. opossum	North American opossum
O. monkey	Owl monkey
P. monkey	Patas monkey
R. deer	Roe deer
R. fox	Red fox
R. hyrax1	Rock hyrax1
R. hyrax2	Rock hyrax2
R. monkey	Rhesus monkey
S. Am. mole	South American mole
S. loris	Slow loris
S.-n. mole	Star-nosed mole
S.-t. shrew	Lesser short-tailed shrew
T. hyrax	Tree hyrax
T. shrew	Tree shrew
W. opossum	Water opossum
Y.-b. marmot	Yellow-bellied marmot

TABLE A.5:   Carcinogen data.

Group	Number of days
1	143, 164, 188, 188, 190, 192, 206, 209, 213, 216, 220, 227, 230, 234, 246, 265, 304, 216+, 244+
2	142, 156, 163, 198, 205, 232, 232, 233, 233, 233, 233, 239, 240, 264, 280, 280, 296, 296, 323, 204+, 344+

TABLE A.6: Clotting
time data.

Lot	Concentration	Time
1	5	118
1	10	58
1	15	42
1	20	35
1	30	27
1	40	25
1	60	21
1	80	19
1	100	18
2	5	69
2	10	35
2	15	26
2	20	21
2	30	18
2	40	16
2	60	13
2	80	12
2	100	12

TABLE A.7: Coal miners data.

Time	Healthy	Normal	Severe
5.8	98	0	0
15.0	51	2	1
21.5	34	6	3
27.5	35	5	8
33.5	32	10	9
39.5	23	7	8
46.0	12	6	10
51.5	4	2	5

TABLE A.8: Fuel consumption data.

Consumption	Temperature	Velocity
14.96	-3.0	15.3
14.10	-1.8	16.4
23.76	-10.0	41.2
13.20	0.7	9.7
18.60	-5.1	19.3
16.79	-6.3	11.4
21.83	-15.5	5.9
16.25	-4.2	24.3
20.98	-8.8	14.7
16.88	-2.3	16.1

TABLE A.9: Gastric cancer data.

Treatment	Time
CR	17, 42, 44, 48, 60, 72, 74, 95, 103, 108, 122, 144, 167, 170, 183, 185, 193, 195, 197, 208, 234, 235, 254, 307, 315, 401, 445, 464, 484, 528, 542, 567, 577, 580, 795, 855, 1174+, 1214, 1232+, 1366, 1455+, 1585+, 1622+, 1626+, 1736+
C	1, 63, 105, 125, 182, 216, 250, 262, 301, 301, 342, 354, 356, 358, 380, 383, 383, 388, 394, 408, 460, 489, 499, 523, 524, 535, 562, 569, 675, 676, 748, 778, 786, 797, 955, 968, 977, 1245, 1271, 1420, 1460+, 1516+, 1551, 1690+, 1694

velocity on fuel consumption are the interesting features in this investigation. The fuel consumption seems to depend at least on air temperature.

Gastric cancer data

A total of 90 patients suffering from gastric cancer were randomly assigned to two groups. One group was treated with chemotherapy and radiation, whereas the other received only chemotherapy (Lindsey 1997, p. 118).

The columns of Table A.9 contain values of treatment and survival time in days. Censored times have been denoted by a + sign. Figure 10.5 on page 244 in Chapter 10 contains a plot of the data.

Insulin data

The effect of insulin on mice, as measured by whether or not they had convulsions was studied (Lindsey 1995a, p. 68).

The columns in Table A.10 contain values of preparation, frequency of convulsions, and total frequency. Figure 6.9 on page 160 in Chapter 6 contains

TABLE A.10:   Insulin data.

Preparation	Dose	Convulsions	Total
Standard	3.4	0	33
Standard	5.2	5	32
Standard	7.0	11	38
Standard	8.5	14	37
Standard	10.5	18	40
Standard	13.0	21	37
Standard	18.0	23	31
Standard	21.0	30	37
Standard	28.0	27	31
Test	6.5	2	40
Test	10.0	10	30
Test	14.0	18	40
Test	21.5	21	35
Test	29.0	27	37

plots showing dependence of observed logits on dose and preparation. There is a clear effect of dose and that effect seems to be the same for both preparations with the standard preparation having higher log-odds than the test preparation.

Leaf springs data

Table A.11 contains free height measurements of leaf springs in the unloaded condition with 8 inches as the target value. The measurements have been done under low and high values of five treatments with three repeats (Pignatello and Ramberg 1985).

The columns in Table A.11 contain values of oil temperature (-/+), transfer (-/+), heating (-/+), furnace (-/+), hold down (-/+), and three height measurements. Figure 3.1 and Figure 3.2 on pages 84 and 85 in Chapter 3 show how factors affect mean and standard deviation of leaf spring height. Factors affecting sample mean include oil temperature, heating time, and furnace temperature. Sample standard deviation is affected by heating time and furnace temperature.

Leukemia data

Data show times of remission (i.e., freedom from symptoms in a precisely defined sense) of leukemia patients, some patients being treated with the drug 6-mercaptopurine (6-MP), the others serving as a control (Cox and Oakes 1984, p. 7).

The columns in Table A.12 contain values of treatment and time of remission. Censored times have been denoted by a + sign. Figure 1.4 on page 12

TABLE A.11:   Leaf springs data.

Oil	Transfer	Heating	Furnace	Hold down	H1	H2	H3
-	-	-	-	-	7.78	7.78	7.81
-	-	-	+	+	8.15	8.18	7.88
-	-	+	-	+	7.50	7.56	7.50
-	-	+	+	-	7.59	7.56	7.75
-	+	-	-	+	7.94	8.00	7.88
-	+	-	+	-	7.69	8.09	8.06
-	+	+	-	-	7.56	7.62	7.44
-	+	+	+	+	7.56	7.81	7.69
+	-	-	-	-	7.50	7.25	7.12
+	-	-	+	+	7.88	7.88	7.44
+	-	+	-	+	7.50	7.56	7.50
+	-	+	+	-	7.63	7.75	7.56
+	+	-	-	+	7.32	7.44	7.44
+	+	-	+	-	7.56	7.69	7.62
+	+	+	-	-	7.18	7.18	7.25
+	+	+	+	+	7.81	7.50	7.59

TABLE A.12:   Leukemia data.

Treatment	Time of remission
Drug	6+, 6, 6, 6, 7, 9+, 10+, 10, 11+, 13, 16, 17+, 19, 20+, 22, 23, 25+, 32+, 32+, 34+, 35+
Control	1, 1, 2, 2, 3, 4, 4, 5, 5, 8, 8, 8, 8, 11, 11, 12, 12, 15, 17, 22, 23

in Chapter 1 plots times of remission. The motive of this investigation is to determine if and how much the drug increases the time to remission.

Lizard data

Table A.13 contains a comparison of site preferences of two species of lizard, *Grahami* and *Opalinus* (Schoener 1970; McCullagh and Nelder 1989, p. 128).

The columns in Table A.13 contain values of illumination, perch diameter in inches, perch height in inches, time of day, species of lizard, and count. Figure 7.1 on page 171 in Chapter 7 shows the dependence of logits on levels of factors.

Malaria data

Proportion of individuals in Bongono, Zaire, with malaria antibodies present was recorded (Morgan, 1992, p. 16; Lindsey 1995a, p. 83).

The columns in Table A.14 contain values of mean age, frequency of seropos-

TABLE A.13: Lizard data.

Illumination	Diameter	Height	Time	Species	Count
Sun	≤ 2	≤ 5	Early	*Grahami*	20
Sun	≤ 2	≤ 5	Early	*Opalinus*	2
Sun	≤ 2	≤ 5	Midday	*Grahami*	8
Sun	≤ 2	≤ 5	Midday	*Opalinus*	1
Sun	≤ 2	≤ 5	Late	*Grahami*	4
Sun	≤ 2	≤ 5	Late	*Opalinus*	4
Sun	≤ 2	≥ 5	Early	*Grahami*	13
Sun	≤ 2	≥ 5	Early	*Opalinus*	0
Sun	≤ 2	≥ 5	Midday	*Grahami*	8
Sun	≤ 2	≥ 5	Midday	*Opalinus*	0
Sun	≤ 2	≥ 5	Late	*Grahami*	12
Sun	≤ 2	≥ 5	Late	*Opalinus*	0
Sun	≥ 2	≤ 5	Early	*Grahami*	8
Sun	≥ 2	≤ 5	Early	*Opalinus*	3
Sun	≥ 2	≤ 5	Midday	*Grahami*	4
Sun	≥ 2	≤ 5	Midday	*Opalinus*	1
Sun	≥ 2	≤ 5	Late	*Grahami*	5
Sun	≥ 2	≤ 5	Late	*Opalinus*	3
Sun	≥ 2	≥ 5	Early	*Grahami*	6
Sun	≥ 2	≥ 5	Early	*Opalinus*	0
Sun	≥ 2	≥ 5	Midday	*Grahami*	0
Sun	≥ 2	≥ 5	Midday	*Opalinus*	0
Sun	≥ 2	≥ 5	Late	*Grahami*	1
Sun	≥ 2	≥ 5	Late	*Opalinus*	1
Shade	≤ 2	≤ 5	Early	*Grahami*	34
Shade	≤ 2	≤ 5	Early	*Opalinus*	11
Shade	≤ 2	≤ 5	Midday	*Grahami*	69
Shade	≤ 2	≤ 5	Midday	*Opalinus*	20
Shade	≤ 2	≤ 5	Late	*Grahami*	18
Shade	≤ 2	≤ 5	Late	*Opalinus*	10
Shade	≤ 2	≥ 5	Early	*Grahami*	31
Shade	≤ 2	≥ 5	Early	*Opalinus*	5
Shade	≤ 2	≥ 5	Midday	*Grahami*	55
Shade	≤ 2	≥ 5	Midday	*Opalinus*	4
Shade	≤ 2	≥ 5	Late	*Grahami*	13
Shade	≤ 2	≥ 5	Late	*Opalinus*	3
Shade	≥ 2	≤ 5	Early	*Grahami*	17
Shade	≥ 2	≤ 5	Early	*Opalinus*	15
Shade	≥ 2	≤ 5	Midday	*Grahami*	60
Shade	≥ 2	≤ 5	Midday	*Opalinus*	32
Shade	≥ 2	≤ 5	Late	*Grahami*	8
Shade	≥ 2	≤ 5	Late	*Opalinus*	8
Shade	≥ 2	≥ 5	Early	*Grahami*	12
Shade	≥ 2	≥ 5	Early	*Opalinus*	1
Shade	≥ 2	≥ 5	Midday	*Grahami*	21
Shade	≥ 2	≥ 5	Midday	*Opalinus*	5
Shade	≥ 2	≥ 5	Late	*Grahami*	4
Shade	≥ 2	≥ 5	Late	*Opalinus*	4

TABLE A.14:   Malaria data.

Mean age	Seropositive	Total
1.0	2	60
2.0	3	63
3.0	3	53
4.0	3	48
5.0	1	31
7.3	18	182
11.9	14	140
17.1	20	138
22.0	20	84
27.5	19	77
32.0	19	58
36.8	24	75
41.6	7	30
49.7	25	62
60.8	44	74

itive, and total frequency. Figure 6.10 on page 161 in Chapter 6 gives a plot of logit against mean age. There is a clear effect of mean age on log-odds.

O-ring data

O-ring thermal distress data contain the number of field-joint O-rings showing thermal distress out of 6, for a launch at the given temperature (°F) and pressure (pounds per square inch) (Dalal *et al.*, 1989).

The columns in Table A.15 contain values of number of thermal distress, temperature (°F), and pressure (psi). Figure 6.1 on page 143 in Chapter 6 shows that proportion seems to depend on temperature but not on pressure.

Pregnancy data

A total of 678 women, who became pregnant under planned pregnancies, were asked how many cycles it took them to get pregnant. The women were classified as smokers and nonsmokers; it is of interest to compare the association between smoking and probability of pregnancy.

The columns in Table A.16 contain values of cycles, smoking habit, and count of women. Figure 9.1 on page 202 of Chapter 9 shows a plot of empirical point probability functions for smokers and nonsmokers. This plot indicates that success probability is higher for nonsmokers than for smokers.

TABLE A.15:   O-ring data.

Count	Temperature	Pressure
0	66	50
1	70	50
0	69	50
0	68	50
0	67	50
0	72	50
0	73	100
0	70	100
1	57	200
1	63	200
1	70	200
0	78	200
0	67	200
2	53	200
0	67	200
0	75	200
0	70	200
0	81	200
0	76	200
0	79	200
2	75	200
0	76	200
1	58	200

TABLE A.16: Pregnancy data.

Cycles	Smoking	Count
1	Smoker	29
1	Nonsmoker	198
2	Smoker	16
2	Nonsmoker	107
3	Smoker	17
3	Nonsmoker	55
4	Smoker	4
4	Nonsmoker	38
5	Smoker	3
5	Nonsmoker	18
6	Smoker	8
6	Nonsmoker	22
7	Smoker	4
7	Nonsmoker	7
8	Smoker	5
8	Nonsmoker	9
9	Smoker	1
9	Nonsmoker	5
10	Smoker	1
10	Nonsmoker	3
11	Smoker	1
11	Nonsmoker	6
12	Smoker	3
12	Nonsmoker	6
>12	Smoker	7
>12	Nonsmoker	12

TABLE A.17:   Puromycin data.

Treatment	Concentration	Velocity
Treated	0.02	76
Treated	0.02	47
Untreated	0.02	67
Untreated	0.02	51
Treated	0.06	97
Treated	0.06	107
Untreated	0.06	84
Untreated	0.06	86
Treated	0.11	123
Treated	0.11	139
Untreated	0.11	98
Untreated	0.11	115
Treated	0.22	159
Treated	0.22	152
Untreated	0.22	131
Untreated	0.22	124
Treated	0.56	191
Treated	0.56	201
Untreated	0.56	144
Untreated	0.56	158
Treated	1.10	207
Treated	1.10	200
Untreated	1.10	160

Puromycin data

The number of counts per minute of radioactive product from an enzymatic reaction was measured as a function of substrate concentration in parts per million and from these counts the initial rate or "velocity" of the reaction was calculated. The experiment was conducted once with the enzyme treated with Puromycin and once with enzyme untreated (Bates and Watts 1988, p. 269). The columns in Table A.17 contain values of treatment, concentration (ppm), and velocity (counts/ min^2). Figure 4.1 on page 109 in Chapter 4 gives a plot of velocity against concentration for treated and untreated groups.

Red clover plants data

Table A.18 contains results of a study of the effect of bacteria on the nitrogen content of red clover plants. The treatment factor is bacteria strain, and it has six levels. Five of the six levels consist of five different *Rhizobium trifolii* bacteria cultures combined with a composite of five *Rhizobium meliloti* strains. The sixth level is a composite of the five *Rhizobium trifolii* strains with the composite of the *Rhizobium meliloti*. Red clover plants are inoculated with the treatments, and nitrogen content is later measured in milligrams. The

TABLE A.18:   Red clover
plants data.

Treatment	Nitrogen content
3DOK1	19.4
3DOK1	32.6
3DOK1	27.0
3DOK1	32.1
3DOK1	33.0
3DOK5	17.7
3DOK5	24.8
3DOK5	27.9
3DOK5	25.2
3DOK5	24.3
3DOK4	17.0
3DOK4	19.4
3DOK4	9.1
3DOK4	11.9
3DOK4	15.8
3DOK7	20.7
3DOK7	21.0
3DOK7	20.5
3DOK7	18.8
3DOK7	18.6
3DOK13	14.3
3DOK13	14.4
3DOK13	11.8
3DOK13	11.6
3DOK13	14.2
COMPOS	17.3
COMPOS	19.4
COMPOS	19.1
COMPOS	16.9
COMPOS	20.8

data are derived from an experiment by Erdman (1946) and are analyzed in Chapters 7 and 8 of Steel and Torrie (1980).

The columns in Table A.18 contain values of treatment factor and nitrogen content. Figure 3.6 on page 99 in Chapter 3 plots nitrogen content values for all treatments. The effect of treatment on nitrogen content is under investigation. The treatment seems to affect the nitrogen content.

Remission of cancer data

The following data set is a set of observations on remission in cancer patients (Agresti 1990, p. 87). The variables in the data set are a labeling index (LI), which measures proliferative activity of cells after a patient receives an

TABLE A.19: Remission
of cancer data.

Index	Total	Remissions
8	2	0
10	2	0
12	3	0
14	3	0
16	3	0
18	1	1
20	3	2
22	2	1
24	1	0
26	1	1
28	1	1
32	1	0
34	1	1
38	3	2

injection of tritiated thymidine. It represents the percentage of cells that are
"labeled." Other variables are the total number of patients with given level
of LI and the number of patients that achieved remission.

The columns in Table A.19 contain values of labeling index, total number
of patients, and number of patients that achieved remission. Figure 1.4 on
page 12 in Chapter 1 shows a plot of observed logits and suggests that the
labeling index has a slightly positive effect on the remission probability.

Respiratory disorders data

Table A.20 summarizes the information from a randomized clinical trial that
compared two treatments (test, placebo) for a respiratory disorder (Stokes *et
al.* 1995, p. 4).

The columns in Table A.20 contain values of treatment, count of favorable
cases, and count of unfavorable cases.

The interest here is how much different the probability of a favorable outcome is in the test group from that of the placebo group. The purpose of
the study was to find out if the *Test* has some effect on the proportion of
unfavorable cases, that is, is there a difference between the proportions of un-

TABLE A.20: Respiratory disorders
data.

Treatment	Favorable	Unfavorable
Placebo	16	48
Test	40	20

TABLE A.21:   Smoking data.

Parental behavior	Nonsmoker	Ex-smoker	Occasional	Regular
Neither	279	79	72	110
Mother	109	29	10	64
Father	192	48	32	131
Both	252	65	32	180

favorable cases in *Placebo* and *Test* groups. There are many ways to express this difference such as the arithmetic difference, the ratio of odds, and the logarithm of the odds ratio for the proportions. For example, the observed difference and the observed logarithm of the odds ratio are -0.417 and -1.792, respectively. These numbers suggest that *Test* has a negative effect on the disorder probability.

Smoking data

This data was obtained from the MRC Derbyshire Smoking Study 1974-1984 (Swan et al. 1991) in which a cohort of more than 6000 children were followed-up through their secondary schooling and subsequently as young adults. The analysis here is to determine how much the smoking behavior of young adult males is associated with parental smoking when they were young children (at age 11-12 years) (Francis et al. 1993).

The columns in Table A.21 contain values of parental smoking behavior and smoking behavior of young adult males for four types of behavior (nonsmoker, ex-smoker, occasional smoker, regular smoker). Figure 2.6 on page 65 in Chapter 2 contains a plot of relative frequencies of different types of smokers for different types of parental behavior and shows that parental behavior somewhat affects propensities to smoke.

Stressful events data

In a study on the relationship between life stresses and illness, one randomly chosen member of each randomly chosen household in a sample from Oakland, California, was interviewed. In a list of 41 events, respondents were asked to note which had occurred within the last 18 months. The results given in Table A.22 are those recalling only one such stressful event (Lindsey 1996, p. 50; Haberman 1978, p. 3).

The columns in Table A.22 contain values of number of months and count. Figure 1.7 on page 18 in Chapter 1 shows plots of count and logarithm of count against month. Here the interest is on the effect of time on recalling stressful events. There is a clear dependence of count on month.

TABLE A.22:
Stressful events
data.

Month	Count
1	15
2	11
3	14
4	17
5	5
6	11
7	10
8	4
9	8
10	10
11	7
12	9
13	11
14	3
15	6
16	1
17	1
18	4

Sulfisoxazole data

In this experiment, sulfisoxazole was administered to a subject intravenously, blood samples were taken at specified times, and the concentration of sulfisoxazole in plasma in micrograms per milliliter (μg/ml) was measured (Bates and Watts 1988, p. 273).

The columns in Table A.23 contain values of time and concentration. The dependence of concentration on time is under investigation. Figure 2.3 on page 58 in Chapter 2 gives a plot of concentration against time. Time has a clear effect on concentration.

Tensile strength data

Data were obtained on the hardness and tensile strengths of 20 samples of an alloy produced under a variety of conditions (Brooks and Dick 1951).

The columns in Table A.24 contain values of hardness and tensile strength. Figure 3.1 on page 84 in Chapter 3 shows a plot of tensile strength against hardness. The effect of hardness on tensile strength is under investigation. There is a clear linear dependence between the variables.

TABLE A.23:
Sulfisoxazole data.

Time	Concentration
0.25	215.6
0.50	189.2
0.75	176.0
1.00	162.8
1.50	138.6
2.00	121.0
3.00	101.2
4.00	88.0
6.00	61.6
12.00	22.0
24.00	4.4
48.00	0.1

TABLE A.24:
Tensile strength data.

Hardness	Strength
52	12.3
54	12.8
56	12.5
57	13.6
60	14.5
61	13.5
62	15.6
64	16.1
66	14.7
68	16.1
69	15.0
70	16.0
71	16.7
71	17.4
73	15.9
76	16.8
76	17.6
77	19.0
80	18.6
83	18.9

TABLE A.25:   Waste data.

Temperature	Environment	Waste output
Low	1	7.09
Low	1	5.90
Low	2	7.94
Low	2	9.15
Low	3	9.23
Low	3	9.85
Low	4	5.43
Low	4	7.73
Low	5	9.43
Low	5	6.90
Medium	1	7.01
Medium	1	5.82
Medium	2	6.18
Medium	2	7.19
Medium	3	7.86
Medium	3	6.33
Medium	4	8.49
Medium	4	8.67
Medium	5	9.62
Medium	5	9.07
High	1	7.78
High	1	7.73
High	2	10.39
High	2	8.78
High	3	9.27
High	3	8.90
High	4	12.17
High	4	10.95
High	5	13.07
High	5	9.76

Waste data

The data are from an experiment designed to study the effect of temperature (at low, medium, and high levels) and environment (five different environments) on waste output in a manufacturing plant (Westfall et al. 1999).

The columns in Table A.25 contain values of temperature, environment, and waste output. The dependence of waste output on temperature and environment is under investigation. Figure 3.7 on page 101 in Chapter 3 shows a plot of values of waste output for different temperatures and environments indicating that both factors might effect the waste output.

Appendix B

Notation Used for Statistical Models

TABLE B.1: Discrete distributions.

Notation	Description
BernoulliModel[ω]	Bernoulli distribution with unknown success probability ω
BetaBinomialModel[n, ω, θ]	Beta-binomial distribution with n trials, unknown success probability ω, and overdispersion θ
BinomialModel[n, ω]	Binomial distribution with n trials and unknown success probability ω
DiscreteModel[k]	Discrete distribution with $1, 2, ..., k$ as outcomes and unknown vector of probabilities as parameter
EmpiricalModel[$\{y_1, \ldots, y_n\}$]	Empirical distribution with numbers y_i as observations
GeometricModel[ω]	Geometric distribution with unknown success probability ω
HypergeometricModel[$n, \omega, n_{\text{tot}}$]	Distribution of the number of successes in a simple random sample of n objects without replacement from a population of n_{tot} objects such that the proportion of successes in the population is ω
LogSeriesModel[ϕ]	Logarithmic series distribution with ϕ as unknown parameter
MultinomialModel[k, m]	k-dimensional multinomial distribution with m trials and unknown probability vector
NegativeBinomialModel[n, ω]	Negative binomial distribution with n trials and unknown success probability ω
PoissonModel[μ]	Poisson distribution with unknown mean μ
ZetaModel[ρ]	Zeta (Zipf-Estoup) distribution with ρ as unknown parameter

TABLE B.2: Continuous distributions.

Notation	Description
BetaModel$[\alpha, \beta]$	Beta distribution with unknown parameters α and β
CauchyModel$[\mu, \sigma]$	Cauchy distribution with unknown location μ and unknown scale σ
ExponentialModel$[\mu]$	Exponential distribution with unknown mean μ
ExtremevalueModel$[\alpha, \beta]$	Extreme value distribution with unknown location α and unknown scale β
GammaModel$[\nu, \mu]$	Gamma distribution with unknown shape ν and unknown mean μ
InverseGaussianModel$[\mu, \tau]$	Inverse Gaussian distribution with unknown mean μ and unknown variance τ
LaplaceModel$[\mu, \sigma]$	Laplace (double exponential) distribution with unknown location μ and unknown scale σ
LogisticModel$[\mu, \sigma]$	Logistic distribution with unknown location μ and unknown scale σ
LogNormalModel$[\mu, \sigma]$	Log-normal distribution such that the natural logarithm of the response has unknown mean μ and unknown standard deviation σ
MultivariateNormalModel$[d, \mu, \Sigma]$	d-dimensional multivariate normal distribution with unknown mean vector μ and unknown variance matrix Σ
NormalModel$[\mu, \sigma]$	Normal distribution with unknown mean μ and unknown standard deviation σ
RayleighModel$[\lambda]$	Rayleigh distribution with unknown scale λ
WeibullModel$[\kappa, \lambda]$	Weibull distribution with unknown shape κ and unknown scale λ

TABLE B.3:   Basic statistical models.

Notation	Description
$\text{IndependenceModel}[\{\mathcal{M}_1, \ldots, \mathcal{M}_k\}]$	Model for statistically independent observations from statistical models $\mathcal{M}_1, \mathcal{M}_2, \ldots$, and $\mathcal{M}_k$
$\text{SamplingModel}[\mathcal{M}, n]$	Model for a sample of size n from statistical model $\mathcal{M}$
$\text{Submodel}[\mathcal{M}, \theta, \omega]$	Submodel of the statistical model $\mathcal{M}$. The expression θ defines the parameter of the submodel and the expression ω gives the parameter of the model $\mathcal{M}$ as function of θ.

TABLE B.4:   Regression models.

Notation	Description
GeneralizedLinearRegressionModel[X, Distribution $\to \mathcal{M}$, LinkFunction $\to g$]	Generalized linear regression model with model matrix X, distribution $\mathcal{M}$, and link function g
LinearRegressionModel[X]	Linear regression model with model matrix X
LogisticRegressionModel[$X, \{m_1, \ldots, m_n\}$]	Logistic regression model with model matrix X and numbers of trials $\{m_1, \ldots, m_n\}$
MultinomialRegressionModel[k, X, $\{m_1, \ldots, m_n\}$]	k-dimensional logistic regression model with model matrix X and numbers of trials $\{m_1, \ldots, m_n\}$
MultivariateLinearRegressionModel[X, d]	d-dimensional multivariate linear regression model with model matrix X
NonlinearRegressionModel[$X, f(x; \beta), x, \beta$]	Nonlinear regression model with model matrix X, regression function f, vector of regressors x, and vector of regression parameters β
PoissonRegressionModel[X]	Poisson regression model with model matrix X
RegressionModel[X, Distribution $\to \{\mathcal{M}_1, \ldots, \mathcal{M}_n\}$, ParameterIndex $\to i$, InverseLink $\to h$]	Generalized regression model with model matrix X, models $\{\mathcal{M}_1, \ldots, \mathcal{M}_n\}$, parameter index i, and inverse link function h
RegressionModel[$X, f(x; \beta), x, \beta$, Distribution $\to \mathcal{M}$, ParameterIndex $\to i$]	Generalized regression model with model matrix X, regression function f, vector of regressors x, vector of regression parameters β, model $\mathcal{M}$, and parameter index i
RegressionModel[$\{X_1, \ldots, X_q\}$, Distribution $\to \{\mathcal{M}_1, \ldots, \mathcal{M}_n\}$, ParameterIndex $\to \{i_1, \ldots, i_q\}$, InverseLink $\to \{h_1, \ldots, h_q\}$]	Generalized regression model with model matrices $\{X_1, \ldots, X_q\}$, models $\{\mathcal{M}_1, \ldots, \mathcal{M}_n\}$, parameter indices $\{i_1, \ldots, i_q\}$, and inverse link functions $\{h_1, \ldots, h_q\}$
RegressionModel[$\{e_1, \ldots, e_n\}, \beta$, Distribution $\to \{\mathcal{M}_1, \ldots, \mathcal{M}_n\}$]	Generalized regression model with expressions e_1, ..., and e_n giving the parameters of the models $\mathcal{M}_1, \mathcal{M}_2, \ldots,$ and $\mathcal{M}_n$ as functions of the regression parameters in β

References

Agresti, A. (1984). *Analysis of Ordinal Categorical Data*. New York: John Wiley and Sons.

Agresti, A. (1990). *Categorical Data Analysis*. New York: John Wiley and Sons.

Akaike, H. (1973). Information theory and an extension of the maximum likelihood principle. In Petrov, B.N. and Csàki, F., *Second International Symposium on Inference Theory*, Budapest: Akadémiai Kiadó, 267–281.

Ashford, J.R. (1959). An approach to an analysis of data for semiquantal responses in biological assay. *Biometrics*, **15**, 573–581.

Azzalini, A. (1996). *Statistical Inference: Based on the Likelihood*. London: Chapman and Hall.

Barndorff-Nielsen, O.E. (1988). *Parametric Statistical Models and Likelihood*. New York: Springer.

Barndorff-Nielsen, O.E. and Cox, D.R. (1989). *Asymptotic Techniques for Use in Statistics*. London: Chapman and Hall.

Barndorff-Nielsen, O.E. and Cox, D.R. (1994). *Inference and Asymptotics*. London: Chapman and Hall.

Bates, D.M. and Watts, D.G. (1988). *Non-linear Regression Analysis & Its Applications*. New York: John Wiley and Sons.

Bernoulli, D. (1777). The most probable choice between several discrepant observations and the formation therefrom of the most likely induction. *Acta Acad. Petrop.*, 3–33. English translation in *Biometrika*, **48**, 3–13 (1961).

Berger, J.O. and Wolpert, R.L. (1984). *The Likelihood Principle*, Lecture Notes – Monograph Series. Hayward, CA: Institute of Mathematical Sciences.

Birnbaum, A. (1962). On the foundations of statistical inference. *Journal of the American Statistical Association*, **57**, 269–306.

Box, G.E.P. (1988). Signal-to-noise ratios, performance criteria, and transformations (with discussion). *Technometrics*, **30**, 1–17.

Brookes, B.C. and Dick, W.F.L. (1951). *Introduction to Statistical Method*. London: Heinemann.

Chen, J. and Jennrich, R.I. (1996). The signed root deviance profile and confidence intervals in maximum likelihood analysis. *Journal of the American Statistical Association*, **91**, 993–998.

Christensen, R. (1990). *Log-Linear Models*. Berlin: Springer.

Clarke, G.Y.P. (1987). Approximate confidence lmits for a parameter function in nonlinear regression. *Journal of the American Statistical Association*, **82**, 221–230.

Collet, D. (1991). *Modelling Binary Data*. London: Chapman and Hall.

Collet, D. (1995). *Modelling Survival Data in Medical Research*. London: Chapman and Hall.

Cook, R.D. (1977). Detection of influential observations in linear regression. *Technometrics*, **19**, 15–18.

Cook, R.D. (1979). Influential observations in linear regression. *Journal of the American Statistical Association*, **74**, 169–174.

Cook, R.D. and Weisberg, S. (1982). *Residuals and Influence in Regression*. London: Chapman and Hall.

Cox, D.R. (1970). *Analysis of Binary Data*. London: Methuen.

Cox, D.R. (1972). Regression models and life tables (with discussion). *Journal of the Royal Statistical Society* B, **34**, 187–220.

Cox, D.R. (1975). Partial likelihood. *Biometrika*, **62**, 269–276.

Cox, D.R. (1988). Some aspects of conditional and asymptotic inference: a review. *Sankhya* A, **50**, 314–337.

Cox, D.R. (1990). Role of models in statistical analysis. *Statistical Science*, **5**, 169–174.

Cox, D.R. and Hinkley, D.V. (1974). *Theoretical Statistics*. London: Chapman and Hall.

Cox, D.R. and Oakes, D. (1984). *Analysis of Survival Data*. London: Chapman and Hall.

Cox, D.R. and Snell, E.J. (1989). *The Analysis of Binary Data*. London: Chapman and Hall.

Dalal, S.R., Fowlkes, E.B., and Hoadley, B. (1989). Risk analysis of the space shuttle: Pre-Challanger prediction of failure. *Journal of the American Statistical Association*, **84**, 945–957.

Dalgaard, P. (2002). *Introductory Statistics with R*. New York: Springer.

Davison, A.C. (2003). *Statistical Models*. Cambridge: Cambridge University Press.

Dobson, A.J. (1990). *An Introduction to Generalized Linear Models*. London: Chapman and Hall.

Draper, N.R. and Smith, H. (1981). *Applied Regression Analysis*. 2nd ed. New York: John Wiley and Sons.

Durbin, J. (1970). On Birnbaum's theorem on the relation between sufficiency, conditionality, and likelihood. *Journal of the American Statistical Association*, **65**, 395–398.

Edwards, A.W.F. (1972). *Likelihood*. Cambridge: Cambridge University Press.

Erdman, L.W. (1946). Studies to determine if antibiosis occurs among Rhizobia. *Journal of the American Society of Agronomy*, **38**, 251–258.

Faraway, J.J. (2005). *Linear Models with R*. Boca Raton, FL: Chapman and Hall/CRC.

Faraway, J.J. (2006). *Extending the Linear Model with R: Generalized Linear, Mixed Effects and Nonparametric Regression Models*. Boca Raton, FL: Chapman and Hall/CRC.

Feigl, P. and Zelen, M. (1965). Estimation of exponential survival probabilities with concomitant information. *Biometrics*, **21**, 826–838.

Fisher, R.A. (1912). On an absolute criterion for fitting frequency curves. *Messenger of Mathematics*, **41**, 155–160.

Fisher, R.A. (1921). On the "probable error" of a coefficient of correlation deduced from a small sample. *Metron*, **1**, 3–32.

Fisher, R.A. (1922). On the mathematical foundations of theoretical statistics. *Philosophical Transactions of the Royal Society*, **222**, 309–368.

Fisher, R.A. (1925). Theory of statistical estimation. *Proceedings of the Cambridge Philosophical Society*, **22**, 700–725.

Fisher, R.A. (1956). *Statistical Methods and Scientific Inference*. Edinburgh: Oliver and Boyd.

Francis, B., Green, M., and Payne, C. (1993). *THE GLIM4 SYSTEM: Release 4 Manual*. Oxford: Clarendon Press.

Fraser, D.A.S. (1968). *The Structure of Inference*. New York: John Wiley and Sons.

Fraser, D.A.S. (1979). *Inference and Linear Models*. New York: McGraw Hill.

Gauss, C.F. (1809). *Theoria Motus Corporum Coelestium in Sectionibus Conicis Solum Ambientium*. Hamburg: Perthes and Besser.

Grimmet, G.R. and Welsh, D.J.A. (1986). *Probability: An Introduction*. Oxford: Clarendon Press.

Haberman, S.J. (1978). *Analysis of Qualitative Data. Volume 1. Introductory Topics*. San Diego, CA: Academic Press.

Hacking, I. (1965). *Logic of Statistical Inference*. Cambridge: Cambridge University Press.

Hurn, M.W., Barker, N.W., and Magath, T.D. (1945). The determination of prothrombin time following the administration of dicumoral with specific reference to thromboplastin. *Journal of Laboratory and Clinical Medicine*, **30**, 432–447.

Jørgensen, B. (1997a). *The Theory of Linear Models*. New York: Chapman and Hall.

Jørgensen, B. (1997b). *The Theory of Dispersion Models*. New York: Chapman and Hall.

Kalbfleisch, J.D. and Prentice, R.I. (1980). *Statistical Analysis of Failure Time Data*. New York: John Wiley and Sons.

Kalbfleisch, J.G. (1985). *Probability and Statistical Inference, Volume 2: Statistical Inference*. New York: Springer.

Knight, K. (2000). *Mathematical Statistics*. New York: Chapman and Hall.

Kullback, S. and Leibler, R.A. (1951). On information and sufficiency. *Annals of Mathematical Statistics*, **22**, 79–86.

Laplace, P.S. (1774). Mémoire sur la probabilité des causes par les évènemens. *Mémoires de l'Académie royale des sciences presents par divers savans*, **6**, 621–656.

Lee, E.T. (1992). *Statistical Methods for Survival Data Analysis*. New York: John Wiley and Sons.

Legendre, A.M. (1805). *Nouvelles Méthodes pour la Detérmination des Orbites des Comètes*. New York: John Wiley and Sons.

Linhart, H. and Zucchini, W. (1986). *Model Selection*. New York: John Wiley and Sons.

Lindsey, J.K. (1974a). Comparison of probability distributions. *Journal of the Royal Statistical Society* B, **36**, 38–47.

Lindsey, J.K. (1974b). Construction and comparison of statistical models. *Journal of the Royal Statistical Society* B, **36**, 418–425.

Lindsey, J.K. (1995a). *Modelling Frequency and Count Data*. Oxford: Clarendon Press.

Lindsey, J.K. (1995b). *Introductory Statistics: The Modelling Approach*. Oxford: Oxford University Press.

Lindsey, J.K. (1996). *Parametric Statistical Inference*. Oxford: Clarendon Press.

Lindsey, J.K. (1997). *Applying Generalized Linear Models*. New York: Springer.

Ludlam, H.A., Nwachukwu, B., Noble, W.C., Swan, A.V., and Phillips, I. (1989). The preservation of micro-organisms in biological specimens stored at $-70°$C. *Journal of Applied Bacteriology*, **67**, 417–423.

McCullagh, P. (1980). Regression models for ordinal data (with discussion). *Journal of the Royal Statistical Society* B, **42**, 109–142.

McCullagh, P. and Nelder, J.A. (1989). *Generalized Linear Models*, 2nd ed. London: Chapman and Hall.

Miller, R.G. (1981). *Survival Analysis*, New York: John Wiley and Sons.

Morgan, B.J.T. (1992). *Analysis of Quantal Response Data*. London: Chapman and Hall.

Nelder, J.A. and Wedderburn, R.W.M. (1972). Generalized linear models. *Journal of the Royal Statistical Society* A, **135**, 370–384.

O'Hagan, A. (1994). *Kendall's Advanced Theory of Statistics, Volume 2B, Bayesian Inference*. London: Edward Arnold.

Pace, L. and Salvan, A. (1997). *Principles of Statistical Inference from a Neo-Fisherian Perspective*. Singapore: World Scientific.

Pawitan, Y. (2001). *In All Likelihood: Statistical Modelling and Inference Using Likelihood*. Oxford: Clarendon Press.

Pignatiello, J.J. and Ramberg, J.S. (1985). Contribution to discussion of offline quality control, parameter design and the Taguchi method. *Journal of Quality Technology*, **17**, 198–206.

Reid, N. (1988). Saddlepoint methods and statistical inference. *Statistical Science*, **3**, 213–227.

Schoener, T.W. (1970). Nonsynchronous spatial overlap of lizards in patchy habitats. *Ecology*, **51**, 408–418.

Searle, S.R. (1971). *Linear Models*. New York: John Wiley and Sons.

Seber, G.A.F. (1977). *Linear Regression Analysis*. New York: John Wiley and Sons.

Sen, A.K. and Srivastava, M.S. (1990). *Regression Analysis, Theory, Methods & Applications*. New York: Springer.

Severini, T.A. (2000). *Likelihood Methods in Statistics*. Oxford: Clarendon Press.

Silvey, S.D. (1980). *Statistical Inference*. London: Chapman and Hall.

Steel, R.G.D. and Torrie, J.H. (1980). *Principles and Procedures of Statistics*. New York: McGraw Hill.

Stigler, S.M. (1986). *The History of Statistics: The Measurement of Uncertainty Before 1900*. Cambridge, MA: Belknap Press.

Stirzaker, D.R. (1994). *Elementary Probability*. Cambridge: Cambridge University Press.

Stokes, M.E., Davis, C.S., and Koch, G.G. (1995). *Categorical Data Analysis: Using the SAS System*. Cary, NC: SAS Institute Inc.

"Student" (1908). The probable error of the mean. *Biometrika*, **6**, 1–25.

Stukel, T.A. (1988). Generalized logistic models. *Journal of the American Statistical Association*, **83**, 426–431.

Swan, A.V., Murray, M., and Jarrett, L. (1991). *Smoking Behaviour from Pre-Adolescence to Young Adulthood*. Aldershot, England: Avebury, Gower Publishing Company.

Tanner, M.A. (1996). *Tools for Statistical Inference: Methods for the Exploration of Posterior Distributions and Likelihood Functions*. 3rd ed. New York: Springer.

Thompson, R. and Baker, R.J. (1981). Composite link functions in generalized linear models. *Applied Statistics*, **30**, 125–131.

Uusipaikka, E.I. (1996). A new method for construction of profile likelihood-based confidence intervals. *1996 Proceedings of the Biometrics Section of the American Statistical Association*, 244–249.

Uusipaikka, E.I. (2006). *Statistical Inference Package SIP.*
  http://www.wolfram.com/products/applications/sip/.
Venzon, D.J. and Moolgavkar, S.H. (1988). A method for computing profile
  likelihood-based confidence intervals. *Applied Statistics,* **37**, 87–94.
Weisberg, S. (1985). *Applied Linear Regression.* 2nd ed. New York: John Wiley and
  Sons.
Welsh, A.H. (1996). *Aspects of Statistical Inference.* New York: John Wiley and
  Sons.
Westfall, P.H., Tobias, R.D., Rom, D., Wolfinger, R.D., and Hochberg, Y. (1999).
  Multiple Comparisons and Multiple Tests: Using the SAS System. Cary, NC: SAS
  Institute Inc.
Wetherill, G.B. (1986). *Regression Analysis with Applications.* London: Chapman
  and Hall.
Wilks, S.S. (1938). The large-sample distribution of the likelihood ratio for testing
  composite hypotheses. *Annals of Mathematical Statistics,* **9**, 60–62.

Data Index

Acute myelogeneous leukemia data, 247, 251

Bacterial concentrations data, 38, 40, 251

Brain weights data, 211, 214, 217, 221, 222, 251

Carcinogen data, 238, 251

Clotting time data, 59, 253

Coal miners data, 66, 198, 253

Fuel consumption data, 6, 21, 24, 27, 36, 41, 42, 49, 51, 84, 89, 91, 95, 97, 98, 103, 226, 253

Gastric cancer data, 243, 257

Insulin data, 161, 257

Leaf springs data, 53, 233, 258

Leukemia data, 12, 29, 34, 44, 46, 245, 258

Lizard data, 169, 184, 259

Malaria data, 165, 259

O-ring data, 142, 152–154, 157, 159, 261

Pregnancy data, 202, 206–208, 261

Puromycin data, 108, 113, 115, 116, 121, 124, 230, 264

Red clover plants data, 98, 264

Remission of cancer data, 61, 148, 149, 163, 265

Respiratory disorders data, 3, 5, 8, 9, 20, 28, 33, 46, 266

Smoking data, 64, 190, 193, 267

Stressful events data, 16, 63, 173, 175, 177, 179, 182, 267

Sulfisoxazole data, 57, 117, 123, 124, 268

Tensile strength data, 83, 89, 91, 103, 268

Waste data, 99, 270

Author Index

"Student", 103

Agresti, A., 61, 199, 265
Akaike, H., 49
Ashford, J.R., 253
Azzalini, A., 49

Baker, R.J., 140
Barker, N.W., 59, 253
Barndorff-Nielsen, O.E., 42, 49
Bates, D.M., xxi, 57, 108, 111, 124,
    264, 268
Berger, J.O., 49
Bernoulli, D., 49
Birnbaum, A., 49
Box, G.E.P., 248
Brookes, B.C., 83, 268

Chen, J., 49
Christensen, R., 187
Clarke, G.Y.P., xxi, 49
Collet, D., 168, 248
Cook, R.D., 79, 103
Cox, D.R., 12, 42, 47, 49, 168, 248,
    249, 258

Dalal, S.R., 142, 261
Dalgaard, P., xxiii
Davis, C.S., 3, 266
Davison, A.C., xxii, 1, 3, 49, 69,
    70, 75, 103
Dick, W.F.L., 83, 268
Dobson, A.J., 140
Draper, N.R., xxvi
Durbin, J., 49

Edwards, A.W.F., 49
Erdman, L.W., 98, 265

Faraway, J.J., xxiii
Feigl, P., 247, 251
Fisher, R.A., 10, 17, 19, 37, 45, 49,
    69, 95
Fowlkes, E.B., 142, 261
Francis, B., 38, 64, 251, 267
Fraser, D.A.S., 49

Gauss, C.F., 81, 82, 103
Green, M., 38, 64, 251, 267
Grimmet, G.R., 49

Haberman, S.J., 16, 187, 267
Hacking, I., 49
Hinkley, D.V., 42, 47, 49
Hoadley, B., 142, 261
Hochberg, Y., 99, 270
Hurn, M.W., 59, 253

Jørgensen, B., xxvi, 140
Jarrett, L., 64
Jennrich, R.I., 49

Kalbfleisch, J.D., 238, 248, 253
Kalbfleisch, J.G., 6, 9, 49, 253
Kass, R.E., 22, 49, 97
Knight, K., 49
Koch, G.G., 3, 266
Kullback, S., 49

Laplace, P.S., 103
Lee, E.T., 248
Legendre, A.M., 87, 103
Leibler, R.A., 49
Lindsey, J.K., xxii, 11, 16, 49, 69,
    70, 161, 165, 187, 243, 247,
    251, 257, 259, 267
Linhart, H., 49

Ludlam, H.A., 38, 251

Magath, T.D., 59, 253
McCullagh, P., 59, 170, 199, 253, 259
McCullagh, P. , 140
Miller, R.G., 248
Moolgavkar, S.H., xxi, 49
Morgan, B.J.T., 165, 168, 259
Murray, M., 64

Nelder, J.A., 59, 69, 140, 170, 253, 259
Noble, W.C., 38, 251
Nwachukwu, B., 38, 251

O'Hagan, A., 49
Oakes, D., 12, 248, 258

Pace, L., 49
Pawitan, Y., 2, 37, 49
Payne, C., 38, 64, 251, 267
Phillips, I., 38, 251
Pignatiello, J.J., 53, 258
Prentice, R.I., 238, 248, 253

Ramberg, J.S., 53, 258
Reid, N., 22, 49, 97
Rom, D., 99, 270

Salvan, A., 49
Schoener, T.W., 170, 259
Searle, S.R., xxvi, 103
Seber, G.A.F., xxvi, 103
Sen, A.K., xxvi, 103

Severini, T.A., 49
Silvey, S.D., 49
Smith, H., xxvi
Snell, E.J., 168
Srivastava, M.S., xxvi, 103
Steel, R.G.D., 98, 265
Stigler, S.M., 50, 81, 82, 104
Stirzaker, D.R., 49
Stokes, M.E., 3, 266
Swan, A.V., 38, 64, 251

Tanner, M.A., 49
Thompson, R., 140
Tobias, R.D., 99, 270
Torrie, J.H., 98, 265

Uusipaikka, E.I., xxii, 33, 49

Venzon, D.J., xxi, 49

Watts, D.G., xxi, 57, 108, 111, 124, 264, 268
Wedderburn, R.W.M., 69, 140
Weisberg, S., xxvi, 103
Welsh, A.H., 49
Welsh, D.J.A., 49
Westfall, P.H., 99, 270
Wetherill, G.B., xxvi
Wilks, S.S., 49
Wolfinger, R.D., 99, 270
Wolpert, R.L., 49

Zelen, M., 247, 251
Zucchini, W., 49

Subject Index

adjusted response, 77, 78, 139
AIC (Akaike's information criterion),
    47, 49, 144, 184
AUC (Area Under Curve), 29, 34,
    120
    confidence interval, 120

Bernoulli distribution
    cumulant generating function,
      144
    moment generating function, 144
BernoulliModel, 243, 271
BetaBinomialModel, 271
BetaModel, 271
binomial distribution, 3, 5, 132, 144
    BinomialModel, 5, 61, 132
    canonical link function, 133
    continuity correction, 145
    cumulant, 144
    cumulant generating function,
      144
    link function, 146
    model function, 5
    moment generating function, 144
    point probability function, 5,
      133
binomial regression, 63, 142
    covariate class, 141
    deviance, 154
    deviance residuals, 157
    LD50, 162
    link function, 142
    log-likelihood function, 151
    logistic regression, 142, 157
    LRT, 154
    MLE, 152
    model checking, 156

modified residuals, 157
    nonlinear, 165
    Pearson residuals, 157
BinomialModel, 5, 61, 133, 141, 148,
    149, 271

canonical link function
    binomial distribution, 133
    gamma distribution, 134
    GLIM, 127
    inverse Gaussian distribution,
      136
    normal distribution, 130
    Poisson distribution, 130
CauchyModel, 148, 149, 271
composite hypothesis, *see* statisti-
    cal inference
confidence interval
    AUC, 31, 120
    calculation, 33
    delta method, 34
    GLM, 98
    Kullback-Leibler divergence, 31
    LD50, 162
    lower limit, 28
    nonlinear regression, 116
    profile likelihood-based, 28
    quantile of normal distribution,
      36
    t-interval, 98
    upper limit, 28
    Wald confidence interval, 35
confidence level, *see* statistical in-
    ference
confidence region
    GLM, 96
    nonlinear regression, 115

continuity correction
  binomial distribution, 145
  negative binomial distribution, 204
  Poisson distribution, 172
Cook statistic
  GRM, 79
covariate class, 141, 189, 201
Cox regression, 240
  hazard function, 240
  partial likelihood function, 242
  risk set, 242
  survivor function, 240
cumulant
  binomial distribution, 144
  gamma distribution, 212
  multinomial distribution, 191
  negative binomial distribution, 203
  Poisson distribution, 170
cumulant generating function
  Bernoulli distribution, 144
  binomial distribution, 144
  gamma distribution, 211
  multinomial distribution, 191
  negative binomial distribution, 203
  Poisson distribution, 170

delta method, *see* confidence interval
density function, *see* statistical model
deviance
  binomial regression, 154
  gamma regression, 221
  negative binomial regression, 208
  Poisson regression, 179
deviance residuals
  binomial regression, 157
  gamma regression, 221
  GRM, 78
  Poisson regression, 182
DiscreteModel, 271

EmpiricalModel, 271

evidential meaning, *see* statistical inference
evidential statement, *see* statistical inference
explanatory variable, 51
ExponentialModel, 29, 271
ExtremevalueModel, 271

F-distribution, 44

gamma distribution, 60, 134
  canonical link function, 134
  cumulant, 212
  cumulant generating function, 211
  density function, 134
  GammaModel, 134
  link function, 213
  moment generating function, 211
gamma regression, 61
  deviance, 221
  deviance residuals, 221
  log-likelihood function, 216
  LRT, 219
  MLE, 216
  model checking, 221
  modified residuals, 222
  Pearson residuals, 222
GammaModel, 60, 134, 214, 271
general linear model, *see* GLM
generalized linear model, *see* GLIM
generalized regression model, *see* GRM
GeneralizedLinearRegressionModel, 214, 271
GeometricModel, 203, 271
GLIM
  adjusted response, 139
  binomial regression, 63, 142
  canonical link function, 127
  gamma regression, 61
  iterative reweighted least squares, 138
  linear predictor, 127
  link function, 127
  log-likelihood function, 129

log-linear model, 169
logistic regression, 63, 142
minimal sufficient statistic, 129
MLE, 136
model function, 127
negative binomial regression, 202
Poisson regression, 64, 169
variance function, 129
GLM, 52, 78, 82
   adjusted response, 78
   confidence interval, 98
   confidence region, 96
   distribution of MLE, 90
   likelihood function, 82
   log-likelihood function, 82
   LRT, 92
   LSE, 87
   MLE, 90
   MLE of variance, 90
   modified residuals, 102
   multiple linear regression model,
      84
   response surface model, 86
   simple linear regression model,
      83
   standardized deviance residu-
      als, 102
   standardized Pearson residuals,
      102
   studentized residuals, 103
   weighted, 225
GRM, 70
   adjusted response, 77
   assumptions, 74
   binomial regression, 142
   Cook statistic, 79
   Cox regression, 75
   deviance residuals, 78
   distinctions, 74
   GLIM, 75
   GLM, 75, 78
   influence, 79
   iterative reweighted least squares,
      76
   leverage, 79

lifetime regression, 75
likelihood ratio statistic, 76
log-likelihood function, 75
log-linear model, 169
logistic regression, 142
MANOVA, 75
many parameter, 71
model matrix, 72
multinomial regression, 75, 190
multivariate, 71
negative binomial regression, 202
nonlinear regression, 75, 78, 107
one-parameter, 71
Poisson regression, 169
proper, 71
quality design model, 231
regression coefficients, 72
regression function, 71
regression parameter, 71
scaled deviance, 76
standardized deviance residuals, 79
standardized Pearson residuals,
   79
Taguchi model, 75, 231
univariate, 71
weighted general linear model,
   226
weighted GLM, 75
weighted nonlinear regression,
   75, 230

hazard function, 240
HypergeometricModel, 271

IndependenceModel, 8, 9, 20, 28,
   29, 34, 39, 44, 271
influence
   GRM, 79
interest function, see statistical model,
   22
inverse Gaussian distribution, 134
   canonical link function, 136
   density function, 136
   InverseGaussianModel, 134

InverseGaussianModel, 136, 271
iterative reweighted least squares,
        76

Kullback-Leibler divergence, 31

LaplaceModel, 271
law of likelihood, 15, 19
LD50, 162
least squares estimate, *see* LSE
leverage
        GRM, 79
lifetime regression, 237
        censoring, 237
likelihood, 10
likelihood approach, *see* statistical
        inference
likelihood function, 4, 10, 11, 23
        GLM, 82
        lifetime regression, 237
        nonlinear regression, 108
        partial likelihood function, 242
        profile, 23
        relative likelihood function, 16
        relative profile, 23
likelihood method, *see* statistical
        inference
likelihood ratio, 15
likelihood ratio statistic, 41
        asymptotic distribution, 42
        GRM, 76
likelihood ratio test, *see* statistical
        inference
likelihood region, *see* statistical in-
        ference
linear predictor
        GLIM, 127
LinearRegressionModel, 44, 84, 86,
        271
link function
        binomial distribution, 146
        binomial regression, 142
        complementary log-log, 146
        gamma distribution, 213
        general binomial, 146

general gamma, 214
general Poisson, 173
GLIM, 127
log, 172
logit, 146
negative binomial regression, 202
Poisson distribution, 172
Poisson regression, 169
probit, 146
reciprocal, 213
log-likelihood function, 11, 23
        binomial regression, 151
        gamma regression, 216
        GLIM, 129
        GLM, 82
        GRM, 75
        lifetime regression, 238
        multinomial regression, 191
        negative binomial regression, 205
        nonlinear regression, 108
        Poisson regression, 176
        profile, 23
        proportional odds regression,
                198
        relative log-likelihood function,
                16
        relative profile, 23
log-linear model, 169, 182
logarithmic likelihood function, *see*
        log-likelihood function
logarithmic profile likelihood func-
        tion, *see* profile log-likelihood
        function
logarithmic relative profile likelihood
        function, *see* relative pro-
        file log-likelihood function
logistic regression, 63, 142, 157, 202
LogisticModel, 271
LogisticRegressionModel, 142, 152,
        271
LogNormalModel, 239, 271
LogSeriesModel, 271
LSE, 87
        GLM, 87
        mean vector, 88

nonlinear regression, 110
normal equations, 88
residuals, 89
variance matrix, 89

maximum likelihood estimate, *see*
    MLE
minimal sufficient statistic
    GLIM, 129
MLE, 11, 47
    asymptotic distribution, 47
    asymptotic normal distribution,
        35
    maximum likelihood estimate,
        45
    nonlinear regression, 112
model function, *see* statistical model
    GLIM, 127
model matrix, 72
model selection
    AIC, 47, 49, 144, 184
modified residuals
    binomial regression, 157
    gamma regression, 222
    GLM, 102
    Poisson regression, 182
moment
    Poisson distribution, 170
moment generating function
    Bernoulli distribution, 144
    binomial distribution, 144
    gamma distribution, 211
    multinomial distribution, 191
    negative binomial distribution,
        203
    Poisson distribution, 170
multinomial distribution, 65, 67, 191
    cumulant, 191
    cumulant generating function,
        191
    moment generating function, 191
multinomial regression, 66, 69, 190
    covariate class, 189
    log-likelihood function, 191
    logistic model, 193

MultinomialModel, 65, 67, 189, 271
MultinomialRegressionModel, 190,
        271
multiple linear regression model, 84
MultivariateLinearRegressionModel,
        271
MultivariateNormalModel, 271

negative binomial distribution, 203
    continuity correction, 204
    cumulant, 203
    cumulant generating function,
        203
    moment generating function, 203
    NegativeBinomialModel, 203
negative binomial regression, 202
    covariate class, 201
    deviance, 208
    link function, 202
    log-likelihood function, 205
    logistic regression, 202, 208
    LRT, 208
    MLE, 206
NegativeBinomialModel, 201, 203,
        271
nonlinear regression, 59, 78, 107
    confidence interval, 116
    confidence region, 115
    distribution of LRT, 115
    distribution of MLE, 112
    likelihood function, 108
    log-likelihood function, 108
    LRT, 121
    LSE, 110
    MLE, 112
    model matrix, 110
    NonlinearRegressionModel, 109,
        120, 121
    normal equations, 110
    standardized deviance residu-
        als, 123
    standardized Pearson residuals,
        124
    studentized residuals, 124
    weighted, 229

NonlinearRegressionModel, 109, 120, 121, 271
normal distribution, 8, 51, 53, 57, 129
    canonical link function, 130
    density function, 130
    NormalModel, 8, 129
normal equations, *see* LSE
NormalModel, 8, 21, 24, 28, 51, 53, 57, 129, 271
nuisance parameter, *see* statistical model

observed information function, 11
observed significance level, *see* statistical inference

parameter, *see* statistical model
parameter space, *see* statistical model
parameter vector, *see* statistical model
parametric statistical model, *see* statistical model
partial likelihood function, 242
Pearson residuals
    binomial regression, 157
    gamma regression, 222
    Poisson regression, 182
point probability function, *see* statistical model
Poisson distribution, 63, 130
    canonical link function, 130
    continuity correction, 172
    cumulant, 170
    cumulant generating function, 170
    link function, 172
    moment, 170
    moment generating function, 170
    point probability function, 130
    PoissonModel, 130
Poisson regression, 64, 169
    deviance, 179
    deviance residuals, 182
    link function, 169
    log-likelihood function, 176

log-linear model, 169, 182
LRT, 179
MLE, 177
model checking, 181
modified residuals, 182
Pearson residuals, 182
PoissonModel, 16, 63, 130, 271
PoissonRegressionModel, 177, 271
probability distribution, *see* statistical model
profile curve, 23
profile likelihood, 23
profile likelihood function, 23
profile likelihood region, *see* statistical inference
profile likelihood-based confidence interval, 28
profile likelihood-based confidence region, *see* statistical inference
profile log-likelihood function, 23
profile trace, 24
proportional hazard regression, *see* Cox regression
proportional odds regression, 69, 195
    log-likelihood function, 198
pure likelihood approach, *see* statistical inference

quality design model, 57, 231

RayleighModel, 271
regression parameter, 51
RegressionModel, 148, 149, 203, 227, 230, 235, 239, 243, 271
regressor variable, 51
relative likelihood function, 16
relative log-likelihood function, 16
relative profile likelihood function, 23
relative profile log-likelihood function, 23
response surface model, 86
response variable, 51
response vector, *see* statistical model

restricted statistical model, *see* statistical inference
risk set, 242
ROC (Receiver Operating Characteristic), 29

sample space, *see* statistical model
SamplingModel, 8, 13, 14, 16, 21, 24, 28, 29, 46, 271
scaled deviance
    GRM, 76
score function, 11
significance level, *see* statistical inference
simple hypothesis, *see* statistical inference
simple linear regression model, 83
standardized deviance residuals
    GLM, 102
    GRM, 79
    nonlinear regression, 123
standardized Pearson residuals
    GLM, 102
    GRM, 79
    nonlinear regression, 124
statistical evidence, 3, *see* statistical inference
statistical hypothesis, *see* statistical inference
statistical inference, 3, 8
    composite hypothesis, 38
    confidence level, 27
    delta method, 34
    evidential meaning, 9
    evidential statement, 9
    interest function, 22
    law of likelihood, 15, 19
    likelihood approach, 9, 20
    likelihood method, 9
    likelihood ratio test, 42
    likelihood region, 19
    LSE, 87
    meaningful statement, 17
    MLE, 90
    observed significance level, 42

profile likelihood region, 26
profile likelihood-based confidence interval, 28
profile likelihood-based confidence region, 27
pure likelihood approach, 19
restricted statistical model, 38
significance level, 42
simple hypothesis, 38
statistical evidence, 3, 9
statistical hypothesis, 38
statistical model, 3
uncertainty, 9
Wald confidence interval, 35
statistical model, 3, 4
    BernoulliModel, 243, 271
    BetaBinomialModel, 271
    BetaModel, 271
    binomial regression, 63
    BinomialModel, 5, 61, 133, 141, 271
    CauchyModel, 271
    Cox regression, 240
    density function, 4
    DiscreteModel, 271
    EmpiricalModel, 271
    ExponentialModel, 29, 271
    ExtremevalueModel, 271
    gamma regression, 61
    GammaModel, 60, 134, 271
    GeneralizedLinearRegressionModel, 271
    GeometricModel, 271
    GLIM, 127
    GLM, 82
    HypergeometricModel, 271
    IndependenceModel, 8, 9, 20, 28, 29, 34, 39, 44, 271
    interest function, 6
    InverseGaussianModel, 136, 271
    LaplaceModel, 271
    lifetime regression, 237
    LinearRegressionModel, 44, 84, 86, 271
    logistic regression, 63

LogisticModel, 271
LogisticRegressionModel, 142,
        152, 271
LogNormalModel, 239, 271
LogSeriesModel, 271
model function, 4
multinomial regression, 66, 69
MultinomialModel, 65, 67, 189,
        271
MultinomialRegressionModel, 190,
        271
MultivariateLinearRegression-
        Model, 271
MultivariateNormalModel, 271
negative binomial logistic re-
        gression, 208
NegativeBinomialModel, 201,
        271
nonlinear regression, 59, 78
NonlinearRegressionModel, 109,
        120, 121, 271
NormalModel, 8, 21, 24, 28,
        51, 53, 57, 129, 271
nuisance parameter, 6
parameter, 4
parameter space, 4
parameter vector, 4
parametric statistical model, 4
point probability function, 4
Poisson regression, 64
PoissonModel, 16, 63, 130, 271
PoissonRegressionModel, 177,
        271
probability distribution, 4
proportional odds regression,
        69, 195
quality design model, 57, 231

RayleighModel, 271
RegressionModel, 227, 230, 235,
        239, 243, 271
response vector, 4
sample space, 4
SamplingModel, 8, 13, 14, 16,
        21, 24, 28, 29, 46, 271
Submodel, 39, 40, 271
Taguchi model, 57, 231
WeibullModel, 13, 14, 271
ZetaModel, 271
studentized residuals
        GLM, 103
        nonlinear regression, 124
Submodel, 40, 271
submodel, 39
survivor function, 240

Taguchi model, 57, 231

uncertainty, see statistical inference

variance function
        GLIM, 129

Wald confidence interval, see con-
        fidence interval
Weibull distribution, 12
        cumulative distribution func-
                tion, 13
        density function, 13
WeibullModel, 13, 14, 271
weighted general linear model, see
        GLM
weighted nonlinear regression, see
        nonlinear regression

ZetaModel, 271

T - #0386 - 071024 - C18 - 234/156/14 - PB - 9780367387082 - Gloss Lamination